BATTLEBOTS®: THE OFFICIAL GUIDE

Mark Clarkson

McGraw-Hill/Osborne
NEW YORK CHICAGO SAN FRANCISCO
LISBON LONDON MADRID MEXICO CITY
MILAN NEW DELHI SAN JUAN
SEOUL SINGAPORE SYDNEY TORONTO

Publisher
Brandon A. Nordin

Vice President &
Associate Publisher
Scott Rogers

Acquisitions Editor
Marjorie McAneny

Project Editor
Jody McKenzie

Acquisitions Coordinator
Tana Diminyatz

Copy Editor
Lunaea Weatherstone

Proofreader
Lisa Theobald

Creative Director
Dodie Shoemaker

Cover Design
Jeff Weeks

Book Design & Art Direction
Ted Holladay
slickted@pacbell.net

Battlebots Photography
Garry Gay
Daniel Longmire
Aengus McGiffin
Mark Jouret

This book was composed
with Adobe InDesign 2.0.

McGraw-Hill/Osborne
2600 Tenth Street
Berkeley, California 94710
U.S.A.

To arrange bulk purchase discounts for sales promotions, premiums, or fund-raisers, please contact McGraw-Hill/Osborne at the above address. For information on translations or book distributors outside the U.S.A., refer to the following:

INTERNATIONAL CONTACT INFORMATION

AUSTRALIA
McGraw-Hill Book Company Australia Pty. Ltd.
TEL +61-2-9417-9899
FAX +61-2-9417-5687
http://www.mcgraw-hill.com.au
books-it_sydney@mcgraw-hill.com

CANADA
McGraw-Hill Ryerson Ltd.
TEL +905-430-5000
FAX +905-430-5020
http://www.mcgrawhill.ca

GREECE, MIDDLE EAST, & AFRICA
(excluding South Africa)
McGraw-Hill Hellas
TEL +30-1-656-0990-3-4
FAX +30-1-654-5525

MEXICO (Also serving Latin America)
McGraw-Hill Interamericana Editores S.A. de C.V.
TEL +525-117-1583
FAX +525-117-1589
http://www.mcgraw-hill.com.mx
fernando_castellanos@mcgraw-hill.com

SINGAPORE (Serving Asia)
McGraw-Hill Book Company
TEL +65-863-1580
FAX +65-862-3354
http://www.mcgraw-hill.com.sg
mghasia@mcgraw-hill.com

SOUTH AFRICA
McGraw-Hill South Africa
TEL +27-11-622-7512
FAX +27-11-622-9045
robyn_swanepoel@mcgraw-hill.com

SPAIN
McGraw-Hill/Interamericana de España, S.A.U.
TEL +34-91-180-3000
FAX +34-91-372-8513
http://www.mcgraw-hill.es
professional@mcgraw-hill.es

UNITED KINGDOM, NORTHERN,
EASTERN, & CENTRAL EUROPE
McGraw-Hill Publishing Company
TEL +44-1-628-502500
FAX +44-1-628-770224
http://www.mcgraw-hill.co.uk
computing_neurope@mcgraw-hill.com

ALL OTHER INQUIRIES Contact:
Osborne/McGraw-Hill
TEL +1-510-549-6600
FAX +1-510-883-7600
http://www.osborne.com
omg_international@mcgraw-hill.com

BattleBots®: The Official Guide

1234567890 WCT WCT 0198765432

ISBN 0-07-222425-8

Bots with saws come.

Bots with claws come.

Bots with saws and claws and jaws come.

Look, sir. Look, sir. Mr. Rozz, sir.

Let's build bots with jaws and saws, sir.

Let's build bots with paws and claws, sir.

First, I'll make a spinning SawBot.

Then I'll make a ramming ClawBot.

You can make a flipping JawBot.

You can make a rip 'em raw Bot.

— From "Bots in Sox"

(With apologies to Dr. Seuss, wherever you are.)

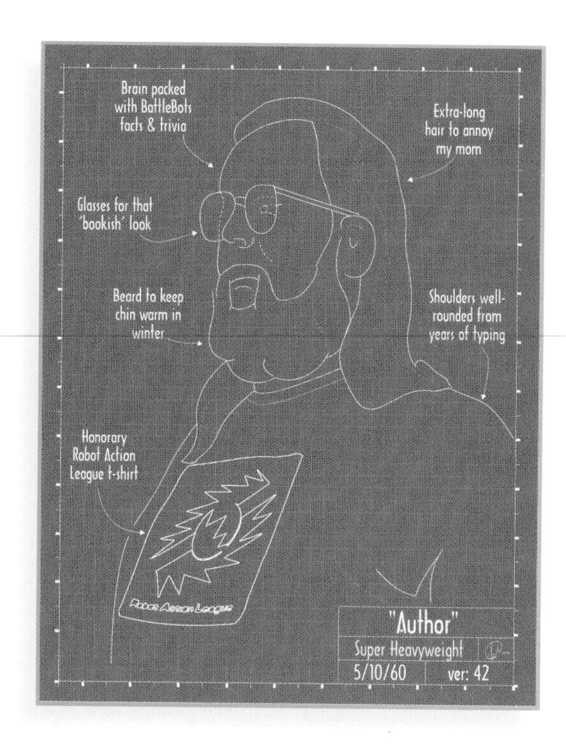

About the Author

Mark Clarkson is a longtime BattleBots enthusiast and technophile. Once a computer programmer and analyst, Clarkson left the world of regular jobs in the 1980s, citing irreconcilable differences. He turned his efforts toward the task of rearranging words into unique and pleasing orders and selling the results. In the 15 years since, he has written about topics ranging from computer-aided BattleBot design to molecular computers and artificial life, to computer games and children's software. In addition to *BattleBots: The Official Guide*, his books include *Flash 5 Cartooning* and *Windows Hothouse: Creating Artificial Life with Visual C++*. At 42, Clarkson, a self-described dilettante, is still trying to decide what to be when he grows up. Until he gets it figured out, Clarkson cartoons, animates, home schools, and writes fact, fiction, and song lyrics from his home office in Wichita, Kansas. Reach out and touch him at www.markclarkson.com.

For Beth,

Who's given me everything

And never told me no.

TABLE OF CONTENTS

BATTLEBOTS

Foreword

BattleBots. Sometimes it takes a while for people to get it. When I explain the concept of BattleBots to them, they might ask, "Are these cartoon robots?" (Yes, there's nothing more exciting than cartoon robot fights.) No, I say, they are real-live fighting robots. Watch it and you'll understand. Then I'll see them a week later, and they'll gush, "That is amazing! I'm hooked! How do those guys make those robots?" That's a little harder for me to explain.

I'm not a technical guy. When a car salesman asks me if I want to take a look under the hood, I hope there's a popcorn maker under there because I won't understand much else. When I'm looking for a sound system, and the guy starts telling me about how many ohms are being pushed, I wonder why those ohms don't stand up for their rights.

But since becoming the color commentator for BattleBots, I have become a bit of a gearhead. Yes, my fingernails are still clean and it's pretty obvious that the only heavy lifting I do is a big box of pens, but now I get it. I understand the beauty of these machines. Sure, you could easily make a machine that could cut through a piece of metal. But can that same machine also withstand the force of a 35-pound hammer over and over? And move around like a wily rabbit? And avoid a titanium spike? Or a gigantic blade? Or a brutal vertical gnasher? That's why they're not just machines. They're BattleBots.

Before, when the subject of robots came up, people would think about a robot that rolls around the living room with a tray of drinks. That robot would get his ass kicked in the BattleBox. And it would be sweet. That robot from *Lost in Space*? He's ground chuck after tussling with Tazbot.

If I'm lost in a desert, give me Tentoumushi to get me through. If I'm living in the Old West, and I've got my back up against the wall and a gang of train-robbing gunslingers are after me, then I want the Judge at my side. Because BattleBots are as tough as they come.

You know, I think even other machines get it. I think that when no one is around the house, the vacuum cleaner and the blender and the humidifier and the can opener and the pencil sharpener and the shoe shiner with the black and red buffers and the garbage disposal all secretly turn on BattleBots and think, "Man, that's what I want to be doing someday."

I know you get it too, or you wouldn't be here holding this book. You're a little twisted, in a good way. You see the splendor in bright purple sparks flying off titanium armor, and the beauty in bulletproof Lexan, freshly painted. You hear the music in a two-cycle chainsaw engine redlining in an enclosed space. You feel the allure of shameless destruction, and respond deep down inside to the dance of two killer machines locked in combat.

This book is for you. Here's your chance to meet these machines, up close and personal. To look under their skirts and see what makes them go. To get to know the people who build the BattleBots, and learn how to build a bot of your own.

Then maybe I'll see you at the next tournament.

—Bil Dwyer

Comedy Central's "BattleBots" co-host and color commentator

Los Angeles, California

March 2002

Acknowledgments

This book would not have been possible without the support of Christian Carlberg, Andrew Lindsey, and Jim Smentowski, fine bot builders all, who went above and beyond the call of duty in helping me meet a very tight schedule. I am indebted to them for much of the information contained here. Any errors or omissions are undoubtedly mine.

Thanks to Trey and Greg for BattleBots, and for helping to make this book happen; to Margie and Jody for helping me keep track of two hundred different things at once; to Katrina for all the files; to Lunaea, the copyeditor, for always knowing where the apostrophe goes; and to Ted for making this book look so fantastic.

Last and not least, thanks to the fine boys and girls on the BattleBots Technical Forum (forums.delphiforums.com/battlebot_tech/) for their enthusiastic input on all subjects; to everyone who took the time and trouble to answer my questions; and to my family for food, love, and support throughout the difficult birthing process, and for still being there when it was over.

Introduction

A Paean to Magnus

Robot combat may be an odd notion to some, but not to me. In my mind, robots and fighting have always gone together. When I was a kid, I loved nothing more than curling up with a stack of Gold Key's "Magnus Robot Fighter 4000AD" comics. (From the sea comes Magnus to fight the evil robots who are the masters of man!)

Wearing his trademark skintight crimson singlet, Magnus watched over the continent-spanning city of North Am, destroying evil battle-robs, giant robot bugs, and other metal baddies with perfectly timed blows of his hands and kicks of his shiny black engineer boots. I can still close my eyes and see Russ Manning's beautifully rendered scenes of combat and destruction—robot bolts and bits sent flying through the air by Magnus' mighty hands and feet. Ah, bliss!

But even Magnus would think twice before kicking Mauler 51-50 once it was up to speed, or before aiming a karate chop at Nightmare's giant spinning disk. His perfect timing wouldn't save him from Toro's super-fast launching arm. His perfect physique wouldn't survive a single whack from Overkill's chrome blade. Magnus may have defeated the deadly undersea Think-Rob, but he wouldn't last ten seconds in the BattleBox.

Yes, kiddies, robot fighting has been made real, but no mere mortal, not even one secretly trained in the martial arts by robot A-1, can stand against the metal monsters you'll find in the BattleBox arena.

A New Sport Is Born

While the robots of Magnus's time generally fought to destroy mankind, the robots in the BattleBox fight for sport.

Is BattleBots really a sport? Damn straight it's a sport. There's no script. There's no fix. It's a straight-up, head-to-head, contest by the rules—albeit by rules that the Marquis of Queensberry never imagined. There are winners and losers. Trophies and acclaim. Cheers and jeers from the stands.

It's not often that one is around to witness the birth of a new sport. Most of them have been around for a long time. Soccer and rugby were invented in the 1820s. Baseball started in 1839, and American football's been around since 1860 or so. Golf dates from the 1400s, basketball was invented relatively recently in 1891, and boxing is roughly as old as walking upright.

In contrast, robot combat only came into being as an organized sport in 1994, less than a decade ago. If you're reading this book, the sport of robotic combat probably isn't even as old as you are. But robot combat is a sport whose time has definitely arrived; it's currently exploding into popularity in the U.S. and across Europe with no end in sight.

BattleBots Tough, Vegetable Nutritious

BattleBots is not just a sport, it's downright gladiatorial. A blood sport. (Well, an oil sport at least.) A fight to the death.

Other sports may be tough, but no other sports are BattleBots tough. Oh sure, you might see a football player pull a muscle or break a leg, but when was the last time you saw one burst into flames? Or held struggling under a sledgehammer by an opponent? In baseball, the batter wears a helmet to protect his head against errant pitches; BattleBots wear titanium armor to protect their guts against carbide-tipped saws. A boxing ring is surrounded by ropes to keep the opponents from falling out and hurting themselves. The BattleBox is surrounded by bulletproof Lexan to prevent supersonic bits of shrapnel from killing people in the audience.

BattleBots is ultra-violent, and yet ... and yet it's as guilt-free a treat as free-range spinach. Hell, it's actually educational. Not only does no one get hurt, but everybody learns something in the process of building and competing: math, science, engineering, welding, and team building. You can't help but learn tons of stuff just by watching the show on TV.

Just Too Rich

The sport of BattleBots is so rich, and so varied, that no book can do more than skim the surface. For every eclectic builder featured in this book, there are dozens more, just as fascinating, just as dedicated to the sport. There are hundreds of BattleBots in active competition, and hundreds more show up for every new tournament. No book can do justice to more than a spare handful. BattleBots comprises way too much coolness to be contained in a few hundred printed pages. Still, someone had to try. And so I have spent all my waking hours for the past few months working on the book you now hold in your hands.

What's in the Book

Though BattleBots is an inherently technical subject, BattleBots: The Official Guide is a fun book (at least, I certainly hope so). The book is highly visual, with hundreds of sexy full-color photographs of your favorite BattleBots and their builders.

History

The book leads off with the short but colorful history of BattleBots and robotic combat. Check it out, even if you hated history in school. You'll find out what German rave parties have to do with robot combat, read about the infamous underground robot rumble of 1998, and hear how Trey Roski was forced, against his will, to found the sport of BattleBots with partner Greg Munson.

The BattleBox

I'll introduce you to Pete Lambertson, the genius engineer behind the BattleBox arena, and we'll step inside the BattleBox to check out the Kill Saws and Pulverizers. We'll visit the pits and run through the safety checklist. You'll read about the judges, the referees, and the CrewBots, and learn about BattleBots rules and regulations, scoring, and ranking.

Who's Doing the Fighting?

You'll read profiles of a dozen of the top teams and builders. Learn who they are and how—and why—they do what they do. Why were Mark Setrakian's disgusting sculptures on display in his high school library? How did Fuzzy Mauldin become a man of leisure? Why is tape considered an illegal weapon in BattleBots? Oh, you'll see.

Types of Robots

Next you'll read about the basic BattleBot types, the Wedges and Spinners and Lifters and Launchers. Find out what distinguishes the different types of BattleBots, and what features and components they all have in common. How is a PoundBot different from a ThwackBot? Why is a Rotary like a writing desk? Are Wedges really for wimps? Find out here.

Robot Profiles

You'll also find profiles of dozens of BattleBots, the big and the small, the silly and the serious. See the robots in all their glory, and learn what they've got under their armor as each robot's builder relates his or her own unique tale. Meet Nightmare. Find out where Sallad got its name. Learn World Peace's secret weakness.

Building Your Own

If you're reading this book in the first place, you might as well admit it: you've already harbored thoughts of building a BattleBot of your very own. No need to deny it. We're all friends here, and we understand. Head on over to Chapter 6, where veteran bot builder and all-around great guy Christian Carlberg will gently guide you through the process of building your very first BattleBot. He'll give you the basic information on what you need and how to put it all together. The rest is up to you.

Winners and BattleStats

Finally, you'll find the skinny on every single robot ever to have competed—name, builder, weight class, win/loss record—and the results of every BattleBots tournament to date. See who won and who lost, who used a hammer, who used a saw, and who used a rotary bludgeoning device.

The next time you get in dispute with the guy in the next cubicle about which robot was the first Super Heavyweight champ, or whether Ziggo whipped Backlash in the fall 2001 quarterfinals, you'll have the information in hand to settle the argument once and for all.

Of course, there's no need to go through this book in order. Just turn to the section you find most interesting and start reading. You can learn more about BattleBots—the sport and the combatants—at www.battlebots.com. You can learn more about me—the man and the dream—at www.markclarkson.com.

BATTLEBOTS®

PART ONE

THE SPORT OF

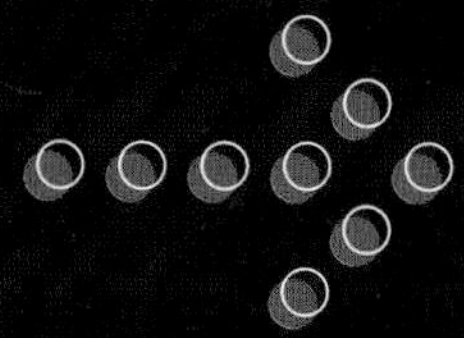

The first remote controlled vehicle was created by Nikola Tesla in 1898. History does not record the first time two remote controlled vehicles were smashed into each other, but it can't have been long afterward. Certainly, boys have been gleefully slamming their remote controlled cars and trucks together for decades. "Isn't that what everyone does?" asks BattleBots cofounder Trey Roski. "Didn't you?"

BattleBots is a remote controlled car combat writ large: kids of all ages gleefully slamming big, handmade killing machines together. Cool? You betcha!

Robotic combat has served as fodder for countless movies, novels, comic books, and daydreams. In the last decade or so, it has become a reality. Today, robot builders across the world compete, *máquina a máquina,* in wide variety of robot sumo matches, soccer matches, climbing contests, and one-on-one and many-on-many robotic fights to the death, both sanctioned and unsanctioned.

ROBOTIC COMBAT

1 > THE HISTORY OF BATTLEBOTS

Greg and Trey's Excellent Adventure

BattleBots cofounders and cousins Greg Munson and Trey Roski never intended to start their own robotic combat sports event, but, in retrospect, their moms can't be too surprised. The two didn't have brothers, so at family get-togethers their moms would thrust them together. The pair managed to get into plenty of trouble, despite the fact that they grew up apart (Munson grew up in the San Francisco Bay Area; Roski grew up in North Hollywood, just down the street from Universal Studios, close enough to hear explosions and car crashes happening daily on the lot). "Growing up," recalls Munson, "Trey and I and our cousin Garner Moss were called the Terrible Trio. We'd always have little destructive, high-tech projects going with remote control cars and airplanes."

Munson also exercised his creative/destructive impulses with Redwood High School friends Mark Setrakian and Peter Abrahamson, who would later become household names among BattleBots fans. "We'd be blowing something up this weekend, building a model the next weekend, making music the weekend after that. We had a band. We'd hang out at each other's houses and watch monster movies."

Setrakian and Abrahamson eventually went on to work in the special effects industry, helping to *create* monster movies. Setrakian met Marc Thorpe at George Lucas's Industrial Light + Magic and, in 1993, Thorpe told Setrakian about this great idea he had for a fighting robot competition, which he called Robot Wars.

BATTLEBOTS®
more history

Building Their Own

Thorpe asked Setrakian to build a combat robot for an upcoming competition. Happy to oblige, Setrakian built a combat robot called The Master, and in 1994 he invited Greg Munson to come to San Francisco to watch The Master compete. "I thought it was great," says Munson. A phone conversation with high school buddy Peter Abrahamson planted the seed for a combat robot of his own.

Team La Ma Motors: Trey Roski, Greg Munson, Garner Moss (not pictured, Gage Cauchois)

And then one night, while taking out the garbage, Munson bumped into downstairs neighbor Gage Cauchois. "I told Gage that I was thinking about building a robot. Gage was interested. He said, 'Let's do it together.'

"I immediately thought of Trey," says Munson. "Trey has great hand-eye coordination skills ... this amazing ability to pilot anything." Munson showed cousin Trey Roski a videotape of Robot Wars and asked him to come onboard. "It was bitchin'!" says Roski. "I said, 'Let's do it!'"

La Machine

The three formed a team, pulled together $600, and set to work creating their robot, La Machine. "Gage did most of the building," says Roski, "but we all put in ideas that made the robot what it was. We built a base that was powered by these funky motors that are used to start remote control boats. It didn't even have a speed controller. We built our own speed controller, but it was really just on and off switches.

"We didn't have much money left, and we were trying to figure out what to do for a weapon," says Roski. "We finally said, hell, let's make a wedge." La Machine is infamous as the very first Wedge in robot competition.

La Machine received a cold reception at the 1995 competition. "We showed up with our Wedge, riveted together out of aircraft aluminum," says Munson, "and people laughed at us. Everybody laughed at us. They made fun of our rivets."

David versus Goliath

"We were right beside a team from U.C. Berkeley," recalls Roski. "They had this very expensive, very sophisticated robot, with five chainsaw motors on it. They looked at La Machine and said, 'What does it *do*? What's the weapon?' But when they went up against us, we destroyed them. That felt real good."

"We were the joke of the event until we started winning," says Munson. "And not only would we win, we'd win in 15 seconds. La Machine was fast, maneuverable, and had a lot of torque. And it really held together well. We hit on the right combination on the first try, and that's a testament to Gage's engineering skills. We were able to get under everybody, throw them off into the side of the arena, and flip them over. And La Machine was a different breed because Trey was driving it. He had this bravado, this finesse and style. It was very exciting to the audience. They loved our robot."

"We showed up with our Wedge, riveted together out of aircraft aluminum, and people laughed at us. They made fun of our rivets."
— Greg Munson

La Machine won its Middleweight weight class and the Middleweight melee. As the Heavyweight melee was about to begin, Roski remembers, the crowd started chanting, "La Machine ... La Machine ... La Machine." Roski says, "Marc Thorpe came up to me and said, 'Do you want to do it? You're gonna get wasted by these heavier robots.' But I said, 'F*@! yeah. We built it to fight! Let's fight!' So we threw it in there, and we beat the crap out of everyone. I did a victory spin, and the crowd just lost it. I was hooked. There's no way I couldn't have been."

Promoting the Sport

"The next stage," says Roski, "was to figure out what we could do to help the sport." Roski helped Marc Thorpe out wherever he could, organizing events, making videos, and spreading the word. He and Munson helped Thorpe make an event tape of the 1995 Robot Wars competition and participated in other promotional schemes.

Before they were bot builders:
Mark Setrakian, Greg Munson, and Paul Rivera, live at Budokan

"We made a fun little promotional tape," recalls Munson. "We took La Machine out on the street and, without having it move, asked people what they thought it was. 'Oh, it looks like that thing you put under an airplane's wheels.' Cut. We show La Machine going under an airplane. 'It looks kind of like a skateboard ramp.' Cut. Guys are jumping over it on skateboards. We convinced the Bay Bridge to open up one of the toll gates for La Machine to drive through. We filmed Marc Thorpe parading around San Francisco with La Machine on a leash."

Munson and Roski also began to appear on TV. "People would hear about the competition and contact Marc Thorpe or the press people," says Munson. "They had a handful of contacts that they sent the media to, and Trey and I were among them.

"We did everything from local morning shows all the way to the national evening news. A couple of British TV shows even came over and interviewed us. We had always sort of fancied ourselves amateur filmmakers, so the fact that we were on TV was tremendous fun for us."

Miniature Rock Stars

Soon they began to travel abroad to promote the sport. "Marc Thorpe's company was contacted by a British production house that wanted to do a pilot for the British version of Robot Wars. There were no British robots, so they flew over a bunch of American robots. It was La Machine, The Master, and Thor. We went over and fought against each other, some prototype British robots, and a French robot. At one point they wanted us to play robot soccer, and La Machine took all the other robots and pushed them into the soccer net."

Munson, Roski, and Scott LaValley, builder of DooAll, even toured Germany on a promotional tour sponsored by Marlboro cigarettes, bringing combat robots to German rave parties. "Everyone would dance until about three in the morning," says Munson. "Then we would go on, and all the glassy-eyed ravers would stand there and stare at these dangerous-looking attack robots as they pushed floaty, billowy weather balloons back and forth. Then we'd do a mock battle at the end. It was a combination of performance art and ass kicking. It was very surreal." They did eleven shows, touring Germany by bus with two DJs and a synth band. "We were in the spotlight of this strange subculture, and we loved it," says Munson. "We felt like miniature rock stars."

The BattleBox crew is ready for anything.

The Rumble Under the Freeway

By 1997, robot combat was becoming embroiled in legal battles. Robot Wars 1998 was cancelled. "There was no outlet for all the builders who had latched on to the sport of robot combat," says Munson, "and everybody wanted to fight."

That "everybody" included father and son team Lowell and Steve Nelson. The Nelsons had built their first robot, S.L.A.M., in 1998 for a competition that never took place. "We were beyond bummed," says Steve Nelson. "We were pissed. We wanted to fight.

"We learned that Jim Smentowski was going to disassemble his heavyweight robot, Hercules, because he was going to give up on the sport. My dad told him, 'You want to take it apart? We'll help you take it apart.' Jim agreed, and we drove [200 miles] from Quincy, California, to Novato, California, so we could kill it for him. There were torrential rains coming down the mountains and washing out the roads. It was a pretty interesting drive."

Smentowski contacted several other area builders, including Roski and Munson, who showed up with La Machine, as well as their new robot, Ginsu. "It was pretty scary watching that thing run the first time," remembers Steve Nelson.

Scott LaValley brought DooAll to the battle, and Steven Felk, who had never competed before, came with his new robot, Voltarc. Smentowski brought his Middleweight robot, Junior. Ironically, Hercules, the robot that started it all, caught fire the night before and was unable to attend. The combatants met at LaValley's house, and then made their way to a freeway overpass where they staged their own fight.

Not wanting to pass up an opportunity to promote the sport, Munson and Roski alerted the news media to their little "underground competition," and a crew from KPIX TV in San Francisco showed up to film the fights.

Should We Go for It?

"At that point," says Munson, "Trey and I were wondering: should we go for it? Should we pick up the torch and create another competition? Trey was passionate, and he knew that if anyone could do it, he could."

Robot combat was mired in lawsuits, but Roski also had access to resources that none of the other robot enthusiasts could muster.

But creating BattleBots was the *last* thing Roski wanted to do. "I never wanted to run the competition," he says. "I'm a competitor. I wanted to compete. I can't compete anymore.

"But nobody else was doing it, and I had the feeling that nobody else was going to do it right. It *had* to be done right. People thought I was nuts—my dad still thinks I'm nuts—but I found a partner, got the money, and we all came up with the name BattleBots."

Of course they couldn't hold all their competitions under freeway overpasses. They needed an arena. Enter Pete Lambertson.

The arena, dubbed the BattleBox, had to look good and provide good visibility for the spectators, but more than that, it had to be safe.

Pete Lambertson Creates the Arena

Munson and Roski met Pete Lambertson when they were competing in Robot Wars. They bought La Machine from Gage Cauchois. Munson says, "Once La Machine did so well, Gage instantly wanted to do something new. He wanted to build Vlad the Impaler. He was driven by it. He was not interested in La Machine any more.

"But we were still interested, and we took La Machine through various incarnations. The second incarnation had a flipper arm, which died in its first match. For the third, Scott LaValley helped us put a pneumatic battering ram on the top of it. The wedge evolved from a flat plane to a scoop. But we couldn't make the scoop in a garage; you need to have industrial bending machines. That's when Trey tracked down Pete."

"I found this place in San Francisco called Standard Sheet Metal," says Roski. "I brought the robot in there and they introduced me to the 'old guy' in the back, the genius engineer, Pete. I instantly fell in love with Pete.

"When we decided we were going to do our own competition, I heard Pete had retired. I called him and said, 'Pete, we need you.' Pete became the arena guy. That's his full-time job, and will be forever."

The arena, dubbed the BattleBox™, had to look good and provide good visibility for the spectators, but more than that, it had to be safe. "We're very concerned with safety," says Roski. "It cost us tons of money to build the BattleBox, but it's safe, and it's getting stronger every year."

The BattleBox was made modular, with 4-foot square floor sections, so it could be assembled in different sizes and configurations. It was built on a raised platform so that weapons and hazards—saws, ramps, and other niceties—could be built under the floor.

"It did a great job," says Roski. "It looked good, and it kept the robots in."

BattleBots at the Pyramid

The first BattleBots competition was held on August 14 and 15, 1999, at the famous Pyramid at Long Beach State University, Roski's alma mater. "In terms of the sport and the builders, it was a huge success," says Munson. "We had about 60 robots show up. All the people who hadn't been able to fight for years now had a chance to try out the bots they'd been working on. They loved it."

At the center of it all stands the BattleBox.

A college friend and independent video producer named John Remar videotaped the event, and Munson and Roski had a 45-minute video edited together. It gave them another tool to use in promoting the sport to investors, television networks, and potential competitors.

The computer-oriented cable channel ZDTV (now TechTV) conducted a live webcast from the event. "The cybercast was very successful," says Munson, "but the audience turnout was less than grand. Most of the spectators were contestant friends and family. I guess in those early days some people just didn't get the concept."

TalentWorks

"A good thing that did happen at the Long Beach event," says Munson, "is that we met the people at TalentWorks, and they took us under their wing." TalentWorks, based in New York City, is a TV production company specializing in sporting events. "They're good people and should be credited with much of our success."

Greg Munson, Trey Roski, and BattleBot Ginsu

TalentWorks helped guide BattleBots through the Byzantine workings of the TV industry. They handle the sports television coverage of the show and are partners with BattleBots in producing it.

"Before TalentWorks," says Munson, "we were being courted by a talent agency who brought us to networks that wanted to change our format, to make it more like a game show or a scripted drama like wrestling. And we were not going to do that at all. We knew it should be positioned like a real sport, and Lenny [Stucker, of TalentWorks] was behind the true sports format. He only considered offers from serious networks that wanted to keep it a sports entity."

"Shortly after the Long Beach show," says Munson, "TalentWorks hooked us up with iN DEMAND pay-per-view, and we planned a pay-per-view event for November 1999."

One new element TalentWorks introduced for the pay-per-view event was a new announcer. "For the Long Beach event," says Roski, "we had this guy from the American Gladiators, Lee Reherman, doing the announcing. He was funny, but we gave him too much leeway. His adlibbing was a little over-the-top."

TalentWorks hired famed ring announcer Mark Beiro to announce the pay-per-view competition. "When Mark Beiro walked in," says Roski, "he didn't know what the hell we were doing. He had this confused look on his face, like 'They're filming robots fighting—what am I doing here?' But after about five matches you could see the change come over him. He started to really get into it.

Mark seemed to put everything in perspective. You could tell he was a keeper."

BattleBots on TV

The second BattleBots competition, filmed for pay-per-view, was held in Las Vegas on November 17, 1999. Along with its new announcer, BattleBots premiered the new Super Heavyweight class. Overall, Roski and Munson were happy with the results. "It was a real, straight-up sports cast, just like we wanted," says Munson. "It went on the air the day before the Super Bowl. But no one watched the damned thing.

"TalentWorks continued taking us around to different television networks," says Munson. "Eventually it was down to ZDTV, now called TechTV, and Comedy Central. We liked ZDTV. They're technology driven. Their attitude toward the sport is ideal. But Comedy Central is basic cable. That means 60–70 million eyeballs. Therein lies the decision."

BattleBots signed a deal with Comedy Central to air the competitions. "We told them that they could have fun with the sport, but they absolutely couldn't make fun of the builders. They assured us that they were not going to make the thing stupid or silly. They were going to make the builders heroes, and at the same time show the world the quirky interesting things they do.

"Season 1.0 of BattleBots was one of the highest rated premieres in their history," says Munson. "Sure, we've had our concerns with some of the things [Comedy Central] does. We're trying to get them to show more of the tournament ladder and more of The Pit. We want to make the show a little more technical. But the bottom line is that the show is great. We're one of the highest rated shows on the network, and we're really happy that the sport has taken off so well."

"BattleBots is true. It's real. There's nothing set up. There's nothing fake."
— Trey Roski

A Growing Sport

The Comedy Central TV show has helped generate a huge groundswell of interest in the sport, and the number of robots and spectators has swollen drastically since the show began to air. "The sport has gotten so big that we've got to split it up into regional competitions," says Munson. "In Season 4.0, we had 400 robots. For Season 5.0, it could be 800 robots. People have to take two weeks off work and fly out to California to compete. That's a lot to ask of someone. We'd rather bring the show as close as we can to them and have fewer people there. That'll give more people more opportunities to enjoy the sport."

Trey Roski, chilling out in the pit area

And it *is* a sport, Munson and Roski insist. "Comedy Central may put a funny spin on the show," says Munson, "but the bottom line is that BattleBots is a straight-up sport."

"BattleBots is true," says Roski. "It's real. There's nothing set up. There's nothing fake. Anyone who shows up with a legal robot that passes the safety inspection can compete, and everyone competes on equal terms.

"It's a boxing match with robots. It's you controlling a robot for a fight to the death where no one gets hurt. There's no way anyone's going to die. You're going to get all your adrenaline [going], but with BattleBots there's no guilt. Is it okay if you wish for that robot to be destroyed? Yes! Wish it! There are no people suffering. The guy whose robot is on fire is smiling and laughing."

Show Me the Money

For BattleBots to continue to evolve as a real sport, Roski and Munson believe it is essential that the competitors be compensated. "The competitors *have* to be paid," says Roski. "We allow sponsorships. Competitors share TV revenue. They get merchandising [benefits]—some of them are going to make a lot of money this year. The guys are putting the money back into their robots, and that makes BattleBots that much better."

But what about the little guy? The garage tinkerer? The retired machinist with a limited budget? Will he still be able to compete in an arena populated by $100,000 BattleBots?

"La Machine cost $600," says Roski. "We beat robots that cost $20,000. Anything can go wrong. A mount can break. A battery can fall off. The greatest thing in the world is when NASA comes in with a robot and they lose to a 14-year-old girl."

BattleBots IQ

"The thing we're most excited about now," says Munson, "is BattleBots IQ." BattleBots IQ is a comprehensive educational program where students learn about the science of engineering through robot building.

"When Trey and I built La Machine with Gage," says Munson, "we had to learn so much. It was not only pulling out the stuff we learned in high school—the algebra and the physics—but budgeting our time and money, promoting ourselves, working together as a team.

"You take a thing that's just an idea in your head and make it into a real, working robot that you can go out and fight against someone elses. Maybe you win and maybe you lose, but that process of creating and building is so valuable, it will empower you in other areas of your life and be an experience that you will treasure for a lifetime.

"And it occurred to us," Munson says, "what a great thing to teach to high school students. First of all, they can apply all the math and physics and science, but they also get to learn the skills that wind up being more important: dealing with people, promoting yourself, budgeting your resources.

It gives them a practical venue to apply what they're learning in school, and it promotes the sport to a young audience. Our motto is 'Mom, BattleBots, and apple pie.'"

Chargin' batteries ...

"Once you've put something together, you can never forget it," says Roski. "What you learn is so much greater that what you could ever lose in competing."

"The goal of BattleBots IQ," says Munson, "is to implement a curriculum that encourages advanced achievement in science and engineering, in a format that captures the imagination of students worldwide, and then to get the thing on TV."

Currently, 17 pilot schools are teaching the BattleBots IQ curriculum, developed by Plymouth, Massachusetts, high school teacher and robot enthusiast Mike Bastoni. Bastoni has been teaching robotics in the classroom for years.

Nola Garcia (of Team Fembot and Team Loki) is in charge of the BattleBots IQ program. Garcia and Bastoni have both participated in FIRST (For Inspiration and Recognition of Science and Technology) robotics competitions with their high school classes. Although they remain big supporters of FIRST, they knew they wanted BattleBots IQ to be different in some fundamental ways.

"In FIRST," says Garcia, "they change the game every year. Every year you have to build a new robot." With BattleBots IQ, teams can enter the same robot, year after year, maintaining it, repairing it, upgrading it, reengineering it, and learning all the skills that go with those activities. To encourange the building of new robots, any robot that wins *too* often will be asked to retire.

"We also wanted a longer building period than [U.S. FIRST's] six weeks," says Garcia. "You cannot take a group of kids, teach them what they need to know, design a robot, build it, and test it in six weeks. In BattleBots IQ, you can take however much time you want, and compete with that same robot as many times you want within the entire school year."

BattleBots IQ tournaments have only one weight class, the equivalent of BattleBot's Middleweight. There are further restrictions on some technologies, such as pneumatics, but the robots are not "dumbed down" for kids; a legal BattleBots IQ robot is essentially a legal BattleBots Middleweight. "That's one thing Greg and Trey were adamant about," says Garcia. "The reason the kids get into the curriculum is that they want to build BattleBots. We've done our best to keep everything the same.

Trey Roski raps with Tonight Show host, Jay Leno, about BattleBots.

"We've built BattleBots IQ on the sports model," says Garcia. "Some of the kids aren't football players or whatever, they're the nerdy kids. This has given them an environment where they find kids like themselves, and they can be on a team together. Just the emotional and social benefits are amazing. One of the parents said, 'Thank you so much! Yes, my son is smart. Yes, he's going to college, but he would have never gone to the prom if he hadn't been on the team with girls.'"

The first BattleBots IQ tournament, for kids 12 to 18 years old, took place on March 29 and 30, 2002, at Universal Studios in Orlando, Florida. More than 40 robots and their teams attended the event, including teams from BattleBots IQ pilot schools.

BattleBots IQ is now spreading beyond the 17 pilot schools. Minnesota is planning to implement the BattleBots IQ program statewide in 2003, in an effort to keep kids interested in manufacturing and keep their manufacturing industry alive.

For BattleBots to continue to evolve as a real sport, Roski and Munson believe it is essential that the competitors be compensated.

Gage Cauchois: Father of the Flying Wedge

Gage Cauchois and Vlad the Impaler

Gage Cauchois has been an avid robot fighter since that fateful day in 1994 when neighbor Greg Munson approached him about building a robot. The result, La Machine, was the first combat wedge. "That high-speed inclined plane was my invention," says Cauchois. "Simple physics. La Machine was the ski jump coming to the skier."

Vlad the Impaler

Cauchois got the idea for his next robot, the legendary Heavyweight Vlad the Impaler, from a videotape of the 1994 Robot Wars: "There was an actual forklift driving around. It didn't work, but I thought, 'Wow, what a concept! Pick the other robot up, then what's he going to do?'"

While in England with Greg, Trey, and La Machine, Cauchois visited a museum featuring historical figures. "They had a ... Vlad the Impaler exhibit," he recalls. "And as soon as I saw that name, it was like an electric spark." Vlad the Impaler, a fast forklift with two fang-like prongs, was born—at least in Cauchois's mind. "I gave it a history," he says. "I created a whole Vlad the Impaler robot mythology, and then just went ahead and built it."

Vlad retired last year, but Cauchois is building its replacement—Vlad the Impaler II—and continues to compete with Super Heavyweight Vladiator.

A Modern Artisan

"I've always been a working artisan," says Cauchois. But his background's not exactly in robotics. For the last 20 years Cauchois specialized in designing and building lighting fixtures. "I'd make a hundred or a thousand of them," he says, "reproducing the same design over and over again.

"With BattleBots, I'm spending more time building one really cool machine; it's more satisfying. And I get the opportunity to work with things I didn't in lighting, like motors." Well, Cauchois did prototype some motorized track lighting once. "That was the closest lighting got to fun," he says. "But the world isn't really ready for motorized lighting."

Just the Bots, Please

For the last year, Cauchois has been building robots exclusively, he says. "I've stopped doing lighting because A.) I was making as much money with the robot and B.) I was sick of lighting.

"Robotics is probably the most multi-disciplinary field there is. That's what I like about it. It's got high technology, machinery, electronics, computers, miniaturization, programming.... You're working in every possible medium.

"Plus you get to travel, and meet Carmen Electra."

Cauchois says that "BattleBots is the ultimate creative toy," but he takes it seriously. "I've never had a robot stop working in a battle. That's why I've done well—my robots don't break.

"If someone's going to build the best robot, it might as well be me."

Pete Lambertson: The Man Behind the BattleBox

"I first met Greg and Trey when they brought their robot, La Machine, in for some work," says Lambertson. At the time, Lambertson worked at San Francisco's Standard Sheet Metal, but he retired soon after.

Pete Lambertson discusses the science of destruction with Science Guy, Bill Nye.

"I was just relaxing and having a good time," Lambertson says. "Playing golf. Then Trey called me up one day and wanted to talk. He had the idea of putting on this robot combat event, and the main thing he was concerned about was the safety of the audience. He asked if I'd be interested in building an arena." At first, Lambertson wasn't sure he was ready to give up his retirement. But ultimately he was hooked: "It sounded like it would be interesting to do, and it would be a challenge, and I said, 'Yeah. I'll do it.'

"We came up with a modular system that you could build in any shape or size," Lambertson says, "and we raised it off the floor so you could put weapons under it." Most of the original arena was fabricated at Standard Sheet Metal, but Lambertson has his own shop now, dedicated to building and improving the BattleBox. Lambertson is in charge of the BattleBox from design and fabrication, through shipping and assembly, to working with the television production company to set up lighting. Oh yes, and he runs the BattleBox during the events, too. Every time you see a Kill Saw emerge from the floor to send a bot flying, or a Pulverizer pound a bot into the arena floor, that's Lambertson at work.

Do people ever feel that they've been treated unfairly by Lambertson's arena? "Oh yes," he says. "I hear that all the time. But I try to be as fair as possible. If I get one robot, I want to get the other robot too.

"We're not there to destroy the robots; we're there to make it more exciting, to throw the robots around. And, of course, the guys who do get hammered on usually make it to TV. We've had people drive their robots into the hammers to make sure they get on television.

"I really enjoy what I'm doing," he says. "It's a lot of hard work, but it's a lot of fun. It's not the kind of work I'm used to; you're in show business. I sign autographs, and that's an experience I never thought I'd have. Trey was always kidding me that he was going to make me famous. I laughed. I thought it was a joke."

"We're not there to destroy the robots, we're there to make it more exciting, to throw the robots around. And, of course, the guys who do get hammered on usually make it to TV."
— Pete Lambertson

Mark Beiro: The Man in the Box

Mark Beiro is one of the top ring announcers in the country. He's been announcing boxing matches since he was 9 years old—professionally since he was 13. "I'm very well known in the boxing community," says Beiro. Is he ever recognized in public? "Yes," he says, "and people usually say: 'Hey! Aren't you that robot guy?'"

BattleBox announcer Mark Beiro wants YOU to build a BattleBot.

Yes, Virginia, Mark Beiro *is* that robot guy, the man who introduces the BattleBots and announces the decisions. In fact, spectators see more of Mark Beiro in the BattleBox than anyone else. In his characteristic ring announcer's tuxedo, with silver hair and sonorous voice, he has become an icon of the burgeoning sport of robotic combat. And no one is more surprised than Beiro himself.

In nearly four decades as a professional, Beiro has announced everything from Jai-Alai to boxing to pro wrestling. He's even been tossed out of the ring by a 350-pound professional wrestler. (All in good fun, he insists; he was unhurt. "Believe it or not," he says, "I'm a great acrobat. But I left in the ambulance and they still had to pay me for the whole night.")

Nothing he'd seen before, however, prepared him for the notion of BattleBots. "I was approached by Lenny Stucker and Robbie Biner of [the New York production company] TalentWorks," says Beiro. "They produce and direct the ESPN2 Friday night fights, and I do a lot of the announcing on those shows. They came up to me at one of the fights and said, 'We're producing this show called BattleBots. It's robots, and they fight each other. It's a new thing, but we really believe in it.' They had the highest enthusiasm.

"I said, 'What do you mean, robots fighting?' And they tried to explain it to me—that there are robots shaped like flying saucers, and some that look like suitcases on wheels with hammers coming out of them—and I just sat there with my mouth open.

"Finally, Lenny says, 'Look, are you going to do it or not?'"

Beiro agreed to announce the event, but when he first stepped into the BattleBox to announce bouts for the Las Vegas 1999 competition, he still didn't know exactly what to expect. "I remember going into the arena and having to explain to people what they were about to see without knowing what they were about to see. I described it as a robot demolition derby. But seeing the first fight cured me. At least now I knew what they were trying to do.

"But I really didn't see the possibilities for the sport," Beiro admits. "I don't think *I* would have gone to see it. But I'll tell you this: you won't ever find me missing one now."

Although he's become a big fan of BattleBots as a sport, Beiro is not himself mechanically inclined at all. "I'm an announcer because I *can't* do this kind of thing," he says. "I don't know a pneumatic from a schematic. Builders are always coming up to me and saying, 'You know, Mark, we put a 347 Jizzo in there.' What are you saying? What are you talking about? I don't even know how to boot a computer."

A BRIEF HISTORY OF ROBOTIC COMBAT

Many thanks to Jim Smentowski, BattleBot's unofficial historian for the information contained in this timeline. For more on the history of robotic combat, visit Jim's site at www.robotcombat.com.

Start

1898 — *Inventor and scientist Nikola Tesla demonstrates the world's first radio controlled vehicle, a boat, at the recently completed Madison Square Garden in New York City.*

Late 1970s — *MUSE Software, an Apple subsidiary, develops a software program called "Robot Wars," whetting the public's appetite for fighting robots.*

1978 — *Mark Pauline founds Survival Research Laboratories, an organization of creative technicians who stage ritualized interactions between machines, robots, and special effects devices as a form of socio-political satire.*

September 1991 — *Dragon*Con science fiction convention adds a Robot Battle to the annual event.*

1992 — *Marc Thorpe fools around in his kitchen attempting to build a radio controlled vacuum cleaner, with unexpectedly violent results. He decides that remote controlled mayhem is a good thing. He starts to plan a robotic combat event.*

1994 — *The first Robot Wars is held, with a handful of competitors. About 1000 people attend to witness the event at the Fort Mason Center in San Francisco.*

2000

January 29, 2000 — *The pay-per-view BattleBots event from November 1999 airs.*

May 6–7, 2000 — *BotBash 2000, similar to BattleBots but with smaller robots, is held in Mesa, Arizona. BattleBots supports the event with BattleBots staff and the use of a quarter-scale BattleBox.*

June 9–11, 2000 — *BattleBots holds its third event to sellout crowds in four sessions over the weekend at San Francisco's Fort Mason Center (back to where it all started). All four weight classes are represented in this single elimination event.*

July 22, 2000 — *Robot Wars U.K. tapes a small show with seven U.S. robots, flown over just for the occasion from the U.S., for future airing on MTV. Some U.K. robots are brought in to join the show.*

August 23, 2000 — *BattleBots TV series begins airing on Comedy Central.*

October 23, 2000 — *BattleBots releases a list of robots entered into the November event — 147 bots in four weight classes.*

October 25, 2000 — *Christian Carlberg and Lisa Winter bring their BattleBots to the Tonight Show with Jay Leno. Jay says he's going to build a bot too.*

November 3, 2000 — *Tonight Show host Jay Leno reveals that he has his own Super Heavyweight robot, Chin-Killa.*

December 12, 2000 — *Season 2.0 of BattleBots begins to air on Comedy Central.*

1995 — The second Robot Wars is held at the Fort Mason Center in San Francisco.

1996 — The third Robot Wars is held at the Fort Mason Center in San Francisco, with 75 robots competing.

1997 — The fourth Robot Wars takes place at San Francisco's Fort Mason Center, with 74 robots competing over the weekend.

Mid–1997 — Robot Wars TV event filmed in the U.K. This series of six episodes airs over six weeks on BBC2 and is a great success.

May 1998 — Carlo Bertocchini, creator of the robot Biohazard, creates an online central discussion area for robot competitors and enthusiasts, now known as the BattleBots Forum.

June 1998 — The Society of Robotic Combat (SORC) forms as a unified group of robotic competitors.

July 1998 — Filming begins on the second series of the Robot Wars U.K. TV show.

March 10, 1999 — BattleBots is announced, founded by Trey Roski and Greg Munson (builders of the famous robot La Machine).

August 14–15, 1999 — The first BattleBots competition is held in Long Beach, California. Around 60 robots attend.

November 17, 1999 — The second BattleBots competition is held in Las Vegas, Nevada. Heavyweights and the new Super Heavyweight class clash at the All American Sports Park for the one day pay-per-view event.

Continue at top

April 2, 2001 — BattleBots closes registration for the May 2001 event. A record-shattering 650+ robots have registered.

April 4, 2001 — TLC begins to air the Robotica series 1.

May 22–28, 2001 — BattleBots Season 3.0 event is held at Treasure Island, San Francisco.

June 20, 2001 — Robot Wars tapes an event in London with several U.S. contestants, to be aired later on TNN.

July 10, 2001 — BattleBots Season 3.0 begins to air on Comedy Central.

August 20, 2001 — The new "Robot Wars Extreme Warriors" series premieres on TNN.

October 5–7, 2001 — BotBash 2001 is held in Phoenix, Arizona.

November 4–11, 2001 — The sixth BattleBots event takes place on Treasure Island, San Francisco, with more than 400 robots in attendance. The event is taped for Comedy Central's BattleBots Season 4.0.

Mid-November, 2001 — Robotica holds another event in Los Angeles, taping the show for TLC's Robotica.

December 13, 2001 — TLC begins to air the Robotica series 2.

January 8, 2002 — BattleBots begins to air Season 4.0 on Comedy Central.

March 29–30, 2002 — The first BattleBots IQ tournament, for kids 12 to 18 years old, takes place at Universal Studios in Orlando, Florida. More than 40 robots and their teams attend the event.

AUTOMATUM TECHNOLOGIES
RYAN
ENGINEERING

BATTLEBOTS

computer
STEW

2 > THE SPORT

I think we all agree that robot combat is a good thing—it's fun, it's educational, it's wholesome family entertainment. But we can't have robots clashing willy-nilly in back alleys, engaging in furtive, catch-as-catch-can encounters. Nobody wants to see robots packing machine guns and incendiary grenades, roaming the streets and shopping malls, eating fire hydrants and launching magazine stands into the air. That's not a sport—that's anarchy!

Fighting robots crave structure. They need a safe place to play, and rules to play by. That's what the sport of BattleBots is here to give them. Locked within the bulletproof BattleBox, BattleBots can attack each other without endangering human lives or arousing the ire of the local constabulary. Playing by agreed-upon rules, they can resolve their differences fairly and openly, without any ugly squabbles about cheating or shouting matches over who actually beat whom.

BattleBots provides a safe environment for bot builders to get their ya-yas out. And BattleBots provides the important infrastructure that makes an activity into a sport: a place to fight, rules to fight by, a crowd to watch and cheer, an announcer to keep the crowd informed, judges to decide close bouts, trophies for the victorious, and fire extinguishers for the combustible.

Here's your chance to learn a little more about the sport of BattleBots: the rules of the game, the role of judges and CrewBots, and the construction of the deadly BattleBox arena. Read on.

BattleBots
Minion
NPC
more on the sport

A BattleBots Tournament

"A BattleBots Tournament," says the official BattleBots literature, "celebrates the sport of robotic combat through a contest of battling machines. Participants design, build, and control BattleBots to demonstrate their creativity, engineering skills, strategy, and driving ability." And that's all true. But a BattleBots tournament is about the people you meet as much as it's about the robots they build.

Complete Control battles Subject to Change Without Reason

The people are the first thing you'll notice when you arrive at a BattleBots tournament. Geeks, nerds, engineers, rocket scientists, mad scientists, mad artists, and amateur mechanics. People pushing robots. People working on robots. People talking about robots. People with a penchant for destruction. People with a special gleam in their eye. People who all share a singular, wacky vision: a vision of building robots for the express purpose of taking other robots apart. In fact, if you're at a BattleBots tournament in the first place, they're people just like you.

"The best part of the competition for me is hanging out with everyone," says Rob Everhart of Team Half-Life. "The fighting is almost secondary. I just wish everyone didn't live so far away so we could hang out more than twice a year."

These people are the reason you go to a BattleBots competition, whether you're a competitor or a fan. Say hi to the builders. Meet them. Ask them about their robots. Make friends. It'll guarantee you a much better time at the competition, and if you end up going head-to-head with these guys in an upcoming battle, you might just be able to borrow a sledgehammer or some extra screws.

Check-in

Your first stop is at the check-in table, where you sign your team in. Your team can be as large or small as you like, but your Pit crew is limited to three (for a Lightweight) to five (for a Super Heavyweight), including yourself. You're allowed the same number of drivers—or operators—as crew members. For safety reasons, team members not part of the Pit crew are not allowed in the Pit, nor are children under eight, the family dog, the family snake, non-team friends and relatives that drop by to say hello, or anyone else without a Pit pass.

Once you've picked up your Pit passes, Pit table assignment, and other paperwork, it's time to unload the robot and head to your Pit table.

The Pit

Look out for that axe!

The Pit area is where you'll get your BattleBot ready for battle, recharge your batteries, and make minor (and major) repairs between matches. The Pit at a BattleBots tournament has to be seen (and heard, and smelled) to really be appreciated. Imagine, if you will, the inside of a large warehouse, festooned with banners. Robots, whole and in

Rob Everhart

Team Half-Life

In Season 2.0, we broke a drive shaft in the semifinals. Team Scrap Daddy and the Revision Z guys all pitched in to help us try to weld the one-inch shaft together so we could give Diesector a run for his money. This great camaraderie is something you can expect in the Pit.

pieces, cover ranks of work tables stretching nearly as far as the eye can see, along with spare robot parts, batteries, battery chargers, radios, electronics, electronic test equipment, candy bar wrappers, metal shavings, duct tape, giant sledgehammers, and other tools of the robotic combat trade. Hundreds of people of all descriptions fill the aisles, attending to their robots. Nearby are special areas for welding and grinding, and for robot testing.

The activity—and the manic enthusiasm of the participants—can be fairly overwhelming. In fact, with hundreds of Pit crews working on hundreds of robots, it's one of the busiest machine shops in the country for the duration of the tournament.

Hey, Friend, Can I Borrow That?

The best thing about the Pit is the people you'll find here. When BattleBox designer Pete Lambertson arrived at his first event at Long Beach in 1999, the thing that impressed him most was the atmosphere in the Pit. "Everybody in the Pit was helping each other," he says. "One guy didn't want to win because the next guy didn't have parts, so he'd help him get the robot ready to compete. It was a great feeling. Everybody seemed to be like a family. That was part of what sold me on the whole business."

In the pits, it's bots and bot parts as far as the eye can see.

"I've been involved in a lot of sporting events," says BattleBots announcer Mark Beiro, "but I've never seen [in any other sport] the kind of camaraderie that exists between BattleBots competitors. It's unrivalled. I've seen people lend parts to people they are going to compete against, so they can compete. That's how sick it is. I'd never seen anything like it."

Everyone in the Pit is quick to help out in your hour of need. They all want to see you compete... even if you end up beating them! "If I destroy somebody's robot with one of mine," say Jim Smentowski of Team Nightmare, "I'm there offering parts and helping to get them back into the arena."

If you forgot to bring a favorite tool, or suddenly need a tool you don't have, you can always count on someone helping out.

The Famous Mauler Hammer

One of the most oft-borrowed tools in the Pit is the notorious Mauler Hammer, an invention of General Knollenberg of the South Bay RoboWarriors, creators of Heavyweight BattleBot Mauler. The 15-pound monster hammer, much heavier than your average sledgehammer, is made from a solid block of steel welded to a steel bar. "It was originally created to knock loose a really big gear," says General Henry Tilford. "The forces involved in the robots are beyond normal hand tools. The Mauler Hammer is the right size to give you the kind of impact you need to bend robots back into shape." Veterans come to ask for it by name, says Tilford. "We also give it to any newbie trying to bend metal with a smaller hammer and getting nowhere. There are no metal presses in the Pit, so instead there is the Mauler Hammer."

For Your Own Protection

Once you're settled into the Pit and have completed those last minute adjustments and assembly, it's time for safety inspection and weigh-in. Expect to spend anywhere from 15 minutes to an hour going through the process.

The purpose of the safety inspection is to make sure that your robot conforms to all the technical regulations and is generally safe to be around... except in the arena, of course.

If your BattleBot spent all of its time locked safely inside the arena, there wouldn't be a need for many safety rules at all, other than those preventing robots with 30 gallons of gas onboard or radioactive armor (and yes, radioactive armor is forbidden).

But in fact BattleBots spend most of their time *outside* the arena, in the Pit, the testing area, or waiting in line for inspection or their next match. And most of the time they spend inside the arena, people are in there with them, moving them into place, getting them ready for the match, shutting them down after a match, and so forth.

A bot's burning in the arena. Call in the CrewBots!

Any time a person is in the proximity of a BattleBot, there's the potential for someone to get hurt. The man in charge of making sure that doesn't happen is BattleBots' chief safety/technical inspector, Frank Jenkins. "We want to make sure the robots are safe to be around," says Jenkins. "We hold people to a fairly high standard of construction engineering. We make sure that anyone building a BattleBot containing pressure bottles, for example, doesn't do things the wrong way."

Jenkins and his team of inspectors work hard to make BattleBots a safe sport, and to make sure that the robots' destructive potential is unleashed only within the locked BattleBox.

"BattleBots has had no serious accidents," says Jenkins, "but you're dealing with things that could kill somebody. I'm in robotics professionally. I like robots. The main reason I do this is that I do not want robots to hurt anybody."

(See the end of this chapter for a condensed safety inspection checklist.)

Safety Inspection

The first item on Jenkins' safety checklist? Safety covers to place over all your robot's sharp points, edges, and corners. These covers must be kept in place at all times, except when you're working on that particular part and, of course, in the BattleBox. "That may sound silly or trivial," says Jenkins, "but most of the accidents that we've had so far have been from things like someone backing into a sharp edge on a BattleBot."

The most common reason for a robot to fail its safety inspection? Failure of the radio transmitter failsafe test. For the test, the BattleBot is mounted on a box with its wheels (or whatever) in the air. With the drive running, the driver shuts off the transmitter. If for some reason your radio control system fails, or if you just turn off the transmitter, your BattleBot must immediately go into safe mode. This means all power must be cut to the drive and weapons systems, shutting the robot down.

"Fifteen to twenty percent of the time," says Jenkins, "they shut off the transmitter and the robot keeps running on. Or worse, something that wasn't running starts running. That's the single biggest item, but almost every item on the safety inspection checklist has been failed at one time or other."

The BattleBots technical regulations continue to change after every tournament. "When you're dealing with a technologically driven sport, especially such a young one," says Jenkins, "it's guaranteed that people are going to be trying things that are totally new, that no one's thought of before.

Flying Bot Brothers — OverKill and Greenspan

"On the one hand, you don't want to restrict creativity, but on the other hand, you've got to make sure that the BattleBots are safe when they're not in combat. There are some people that are amazingly... naïve is the nicest term I can come up with. They simply don't seem to be aware that their BattleBot can injure them or someone else."

Any time Jenkins or his team see anything they don't like, whether it's prohibited by the regulations or not, they can nix your robot. "If we determine that it's not safe, we won't let it run," he says. If that happens, you can either fix the problem to the safety inspector's satisfaction, or pack up your robot and go find a good seat to watch the rest of the tournament.

Gaylan Douglas

Team Xtremebots

We had a serious mishap in the staging line. We tried to power up Phere and got nothing. Zilch, zippo, nada! We frantically started checking everything, which meant pulling off the dome to perform the search. As luck would have it, Fuzzy (Michael Mauldin of Team Toad) was on hand. Fuzzy, through his wisdom and clear thinking, realized that Phere was not receiving a radio signal. So off Fuzzy runs to get his transmitter module and spare receiver crystal as we pull the wedge off to get at the receiver module. Of course, all this was being done under the all-seeing Comedy Central cameras… just great! Because of the great help from our bot mentor, Fuzzy, and the graciousness of Team OJ, we got into the box.

Waiting to Go On

As your match time approaches, you are escorted from the Pit to the queuing area and, finally, to the pre-match staging area just outside the arena, called the Battle Box. Now the adrenaline really starts pumping, and, to make matters worse, TV crews are pointing cameras at you and doing pre-match interviews. Finally, it's time to enter the arena.

The tournament officials responsible for escorting you and your BattleBot into and out of the arena, before and after the match, are dubbed CrewBots. The CrewBots supervise Pit crew members in activating and deactivating the robots. They lock the box before the match begins, and they're the only ones who can unlock it again when the match is over. If something happens that necessitates stopping the match—a bot bursting into flame, perhaps, or two bots getting stuck together—it is the fearless CrewBots who charge in with fire extinguishers and pry bars to set the situation right again.

Referees waiting patiently between bouts in the BattleBox.

Your CrewBots usher you to your place in the arena, and they supervise while you take the covers and restraints off all your robot's pointy, grabby bits. The ring announcer enters and announces the robots. The crowd cheers (and boos). Then it's time to activate your robot, cross your fingers, and head to your driving platform, outside the BattleBox.

It's robot fighting time!

The Bout

A bout is a match between two BattleBots (although either or both of the BattleBots could be a MultiBot, a BattleBot composed of multiple, separate robots or "segments"). In many ways, a bout is similar to a boxing match. Two competitors start at opposite sides of the arena. When the starting buzzer sounds, they have at each other for three minutes or until one of the robots is knocked out.

Two eight-foot by eight-foot squares are painted on opposite sides of the BattleBox floor, one red and one blue. Red and blue sides are assigned at random. There is no particular advantage or disadvantage to either side.

Each robot starts the match on its own colored square. Your robot must fit entirely within its square at the start of the bout, although once the bout begins it can unfurl into a larger configuration or split into separate components.

No, they're not real estate salesmen. These red-jacketed officials are the judges.

Your BattleBot must remain motionless until the match starts. Engines, motors, pumps, and so forth can be running—at idle—but there can be no weapon movement and no movement up, down, or across the arena floor. If either robot does move prematurely, a referee will declare a fault and the match must be restarted. Three faults will cost you the match.

There are two referees, one for each team. Once the BattleBox is cleared and locked, the referee will make certain that you are ready to begin the match, and then press a button that activates half of the countdown switch. When both buttons have been pressed, a three-second countdown begins, visible on the light-tree. When the green light comes on, a buzzer sounds and the match is underway.

Skid Mark up against Backlash

Tom Petrucelli

Team Attitude

Turtle Road Kill is an old timer no longer competitive for a myriad of reasons. He was slated for retirement after Season 4.0. [Teammate] Tom Steyer had entered him, but a family emergency prevented Tom from showing. I was on the verge of forfeiting, when Jim Smentowski (Team Nightmare) intervened saying Turtle … should be allowed to "go down fighting." I acquiesced; Jim drove Turtle and won the next three fights, one step away from the quarters! Jim's love of robots and subsequent great driving allowed Turtle to retire with dignity!

Does This Look Incapacitated to You?

To continue the bout, a BattleBot must be capable of controlled "translational movement." In other words, the driver must be able to move the BattleBot around on the arena floor, at least a little bit. Weapon movement doesn't count: a saw blade might still be spinning, for example, but if the robot can't move from place to place, it's out of the fight.

Village Idiot rides the spikes.

A robot that can't move for 30 seconds is considered incapacitated. This includes robots stuck on some part of the arena—jammed up on the spike strips, for instance. If a referee suspects that your BattleBot is incapacitated, he or she will ask you to prove that it can still move. If you're unable to move, the referee begins a 30-second count, the last 10 seconds of which are verbally counted down. If the count reaches zero before your robot moves, you lose. If your robot moves at any point during the countdown, the countdown stops and the match continues.

You can, at any time, "tap out," voluntarily ending the bout and taking a loss. If your robot is stuck under the Pulverizer, for instance, you might prefer to tap out rather than incur 30 more seconds of vicious pounding, and the damage that will result, before the referee counts you out.

Stuck on Each Other

Your robot is allowed to grab opponents, to lift them (in whole or in part) off the arena floor, and to pin them against the wall—but only for 30 seconds at a time. By the end of that 30 seconds, you must release your opponent... if you can. Failure to release the other robot voluntarily will get you disqualified and cost you the match.

Shaft takes the floor.

At times, however, robots get hung up on each other—my sword jammed in your tank treads, maybe—and are unable to separate again. If this happens, the referees declare a timeout. The robots are deactivated, the BattleBox is unlocked, and CrewBots and Pit crews enter the box, armed with crowbars or other appropriate tools of the trade, to pry the combatants apart. When the robots have been separated, and the arena is cleared and locked again, and the bout resumes where it left off. The timer is not reset.

Judges doing their stuff at arena side.

Technically, I Knocked You Out

If your opponent's BattleBot becomes incapacitated (incapable of moving around the arena) while yours can still move, you win. Congratulations!

If your robot personally incapacitated its opponent, or at least had a hand in it, it scores a knockout (KO). You needn't have destroyed an opponent with your weapon to score a knockout; you might have delivered it to a Pulverizer or the Kill Saws. Or just slammed into it hard enough to jar something loose.

If your opponent takes itself out of action without any help whatsoever from you, it's still a win, but it's scored as a technical knockout (TKO). Both KOs and TKOs earn your bot two points.

If both robots become incapacitated before the end of the bout, the win is awarded to whichever robot was the last one moving.

The Judgment and Verdict

If the bout goes the entire three minutes without either bot being incapacitated, the outcome is decided by a panel of three judges seated at arena side. Judges score each match in three categories: aggression, damage, and strategy.

- **Aggression** is a measure of how often you attacked your opponent, how boldly, and how severely.
- **Damage** is just what it sounds like: how much damage you inflicted on your opponent, either directly with your robot or weapons, or indirectly with the arena hazards. A weaponless robot that successfully and consistently pushes its opponents under the Pulverizers can outscore robots with ferocious weapons.
- **Strategy** measures how successfully you carry out a game plan that exploits your robot's strengths against your opponent's weaknesses, while protecting your robot's weaknesses against your opponent's strengths. (Note: Running away does not count as a strategy.)

Each judge divides a total of 5 points between the two BattleBots in each category. A hypothetical perfect score, then, would be 45 points: 5 points in each of the three categories, from each of the three judges. Note that, since 45 does not divide evenly by 2, there can never be a tie in the judges' decision.

The judges' scorecards determine a bot's fate.

Of course, all the strategy and aggression in the world won't do you any good if your robot breaks before the three-minute bout is over.

CompuBot

In addition to the judges' scores, BattleBots generates a CompuBot score for each bout. CompuBot statistics include hits, flips, pins, escapes, weapons damage, and hazard damage. Like the CompuBox system in boxing, which records statistics such as uppercuts and body blows, CompuBot is primarily for fun. The results are unofficial; CompuBot is completely separate from the criteria used in the official judging, and the judges never see the CompuBot statistics until after the bout. Comedy Central generally shows parts of the CompuBot statistics at the end of any bout that goes to a judge's decision.

A Rumble in the Arena

BattleBots tournaments feature a second type of match called the Robot Rumble. The rules of the Rumble are similar to the rules of the standard bout, with a few exceptions.

In a Rumble, up to 18 robots compete to be the last bot standing. As there are only two colored squares, at the beginning of the Rumble, combatants are spread evenly

Derek Young

Automatum Technologies

Before Complete Control's first match ever, I set the robot down for the final test run to make sure everything was operational. The robot slowly crept forward and then just sat there. Panic ensued. I tested the motors. It appeared that the brushes were not making contact with the commutator of the motor. I quickly pulled one of the brushes and found that the brush springs appeared to have annealed and shrunk, pulling the brush away from the commutator. At this point, since I didn't have spares, I was ready to give up. But wait! Charles Carns suggested we replace the brush springs with ballpoint pen springs! We proceeded to bust apart a bunch of pens and, with three soldering irons and five pairs of hands, got the motors back together and running minutes before the match. Big thanks to Charles and Team Carnivore for helping me out when I thought I was totally done.

around the outside of the BattleBox. Rumbles are longer than bouts, lasting five minutes rather than three. You cannot tap-out in a Rumble, and Rumbles are not stopped if robots become stuck together.

There are two Rumbles for each weight class: the Main Rumble and the Consolation Rumble. The Main Rumble is for the top 16 of the 32 finalists in each weight class. The Consolation Rumble is for the bottom 16. The Consolation Rumble is held first, and up to two winners from the Consolation Rumble are also eligible to compete in the Main Rumble.

The winner of a Rumble is determined in the same way as the winner of a bout. If only one robot is still moving at the end of the five-minute match, it wins. If more than one robot is still responsive, the winner is determined by a judge's decision.

Single Elimination

BattleBots competitions are single-elimination tournaments: the first time you lose a match, your robot is out of the competition. As long as you keep winning matches, you keep moving up to the next round. The exact number of rounds varies from weight class to weight class, and from competition to competition, depending on the number of robots competing. But build your bot strong—it will have to survive six or more punishing matches, against increasingly powerful foes, if you plan to take home the coveted Giant Nut.

The coveted Giant Nut.

Ranking Points

Points are awarded for every win in a BattleBots tournament, except for the Consolation Rumble. If you win, you'll receive the following:

- **Two points** for each win in any single-elimination tournament match.
- **One point** for each loser's bracket win in any double-elimination match. (Note: BattleBots competitions are currently all single-elimination tournaments.)
- **One point** for each win caused by opponent forfeit.
- **Two points** for each Main Rumble win.
- **Two points** for each tournament win.
- **One point** for the tournament runner-up.

You'll also get a nifty win pog!

The more points your BattleBot wins, the higher its ranking. The robot with the most points in its weight class is the #1 ranked BattleBot in that class.

This robot is not necessarily the current champion, however. BattleBots uses an active seeding system; the win percentage earned for the last three tournaments is used to determine your current active seeding. BattleBots keeps historical records, which can be found at the end of this book. BattleBots also tracks such stats as knockouts, average knockout time, and total judges' points. These stats come into play if there's a tie in points between two robots.

Weight Class Designation

	More Than	Maximum
Lightweight	25.0 lbs.	60.0 lbs.
Middleweight	60.0 lbs.	120.0 lbs.
Heavyweight	120.0 lbs.	220.0 lbs.
Super Heavyweight	220.0 lbs.	340.0 lbs.

The BattleBox

All BattleBots bouts take place within a completely enclosed arena called the BattleBox. The number one job of the BattleBox is to contain the robots and any bits that may come flying off them during a match. To that end, the arena floor is made from quarter-inch thick steel, and the walls and ceiling are made of bullet-proof Lexan.

Modular Design

The BattleBox is modular; everything breaks down into four-foot by four-foot sections, which are easy to pack and ship. The BattleBox's 2300-square-foot steel floor is actually an assembly of those sections, each of which stands two feet high on steel tube framing. The flooring modules bolt and clip together into one solid unit. The BattleBots arena, as seen on TV, consists of 144 floor modules arrayed in a 12 by 12 square (48 feet by 48 feet). But it can be assembled in other sizes and configurations. When the BattleBox appeared on the Tonight Show with Jay Leno, for example, it came in its handy compact traveling size of six by six modules. The floor weapons are also built into interchangeable four-foot-square modules. This means the configuration of the arena—the number and placement of weapons and hazards—can vary.

The BattleBox is bigger and better decorated than my first house.

Lexan walls drop into channels, or "ears," in the outermost flooring modules. The walls (also modular) vary in thickness by sections, from an inch thick at the bottom to a quarter inch thick at the top, and reach a height of 16 feet above the arena floor.

The original BattleBox had a sheet of flexible plastic across the top to stop any stray bot parts flying that high. But stronger, more destructive robots, especially vertical SpinBots such as Nightmare, forced BattleBots to add a quarter-inch Lexan lid to the arena in 2001.

The Weapons

The original BattleBox had two saws and two spike units in the floor. Well-known BattleBots builder Christian Carlberg built those first weapons. Today the BattleBox is bristling with hazards: saws, spinning turntables, pistons, spikes, heavy pneumatic sledgehammers, and whatever BattleBox designer and operator Pete Lambertson thinks of next.

One of the feared Kill Saws, in a moment of repose.

It's important to note, for those planning on using the arena hazards extensively, that the hazards don't come into play until the final rounds of a tournament; the early rounds take place without them. Even in the later rounds, insists Lambertson, the purpose of the arena hazards isn't to destroy the robots. The hazards are there "to make it more exciting," he says. "Some of the early rounds without the hazards can get a little dull when it's just two boxes pushing each other around, like bumper cars."

One of the BattleBox's revolving screws.

Carlo Bertocchini
Team Biohazard

In Season 4.0, Jabberwock's air-powered battering ram got Biohazard just right and broke one of the pieces on the arm. The competition isn't too far from my home, so I actually took the whole thing home, worked all night on it, and made a new part. It would normally have taken me about two days, but I did it in a rush in about eight hours. The very first fight the next day I went against Nightmare. ... He hit me on the very same part I'd spent all night working on and broke it again.

Tazbot competes against GoldDigger

Techno Destructo vents a cloud of CO_2

Jim Smentowski

Team Nightmare

If you keep on winning, toward the end of the tournament they only have to give you a minimum of 20 minutes [between matches]. Back in Season 2.0, Backlash went all the way to the finals, and Nightmare made it to the quarterfinals. I was totally stressed, in and out of the arena, back and forth, back and forth. I didn't even have time to let the batteries get cooled down, let alone recharge them.

Take the Kill Saws, for example. They wield four sets of 20-inch, carbide-tipped, SystiMatic saw blades, spun by a powerful 5 HP electric motor. "But we don't actually want to cut the robots," says Lambertson. "We want to throw them. I worked with SystiMatic and they helped me figure out which blades to use to get the effect that we wanted. The Kill Saws use three saw blades now, sandwiched together. They will cut the robot just a little and throw it at the same time. It's more exciting."

The stainless steel head of a Pulverizer weighs a whopping 35 pounds.

The Pulverizers have evolved from store-bought sledgehammers with plastic handles to super-tough, custom-made stainless steel monsters with 35-pound heads, but those heads are hollow. Again, says Lambertson, the Pulverizer isn't intended to destroy robots. "But we will put dents in them!" he says. "If we really wanted to smash the robots, we'd go with a solid hammer head."

The Pulverizers are driven by pneumatics, as are most of the arena's weapons. Everything is run from Lambertson's big control board at arena side.

The crowd loves it when something gets destroyed in the Box.

Rearranging the Arena

Thanks to its modular design, the floor of the arena can and does vary from tournament to tournament. Although the size remains the same, weapons are always being added, subtracted, modified, or moved to new positions. "We're always trying to create something new," says Lambertson, "so you don't see exactly the same thing every time. It keeps all the robot builders on their toes, too, because they don't know what I'm going to come up with next."

For Season 4.0 in November 2001, for example, the number of hazards was reduced. "We wanted to give the robots a little more open space in the middle of the arena," says Lambertson.

Some hazards have disappeared from the arena altogether, such as ramps that popped out of the floor. "The ramps were put in originally for the robots to jump over," says Lambertson. But no one used them for that, and the ramps served mainly to tip robots over. The ramps are gone, at least for now.

"It's a process of creating as we go along to make things more exciting for the audience," says Lambertson.

CrewBots working on one of the BattleBox's Kill Saw units.

Safety/Technical Inspection Checklist

This is a condensed version of the Safety/Technical Inspection Checklist from the BattleBots Tournament Rules & Procedures. For a complete, up-to-date checklist, download the Tournament Rules & Procedures from the BattleBots Web site.

South Bay RoboWarriors give Mauler a thorough check.

Team Blendo gets under the hood.

> External Inspection

- ❍ All sharp points, edges, and corners have safety covers.
- ❍ All pinch and motion hazards (such as jaws) are restrained.

> General Internal Inspection

- ❍ Master switches comply with the technical regulations.
- ❍ All batteries are of the proper type and securely mounted.
- ❍ Wiring is properly installed, and electrical terminals are covered and insulated.

> Pneumatics

- ❍ All components are properly rated and tested, and properly secured.
- ❍ The system is properly designed and constructed.
- ❍ The system has all appropriate pressure-reliefs, and shut-off and purge valves.
- ❍ Inspector observes the pressurization procedure and the condition of filling tanks and hardware.
- ❍ When completely filled, all pressures are within limits.

> Hydraulic Systems

- ❍ All hydraulic components are properly rated and securely mounted.
- ❍ All hoses are properly supported.
- ❍ The system has appropriate pressure reliefs.
- ❍ Hydraulic fluid won't spill out if the BattleBot is flipped upside-down.

> The Weigh-in

- ❍ All tanks are full and pressurized, all fuel is onboard, and the BattleBot is in battle-ready condition.

Ian Watts

Team Big Brother

For our first fight, we drew up against Mauler. Everyone came to us and said, in funereal tones, "Sorry, mate. Bad luck on the draw." That evening we went to the bar and drank far too much Budweiser, drowning our sorrows and hatching plans to run away.

The next morning saw us very hung over and feeling poorly when one of the crew came up and said, "I've got this piece of arena wall here, any use to you?" We thanked him profusely, and we cut it up and covered the robot with it. About 10 minutes after we finished, another member of the crew was walking about saying, "Has anyone seen that bit of Lexan I cut specially for the Kill Saws?"

I'm sorry to say we kept quiet about where it went, and we'd like to say that we're really sorry. To cap it all, we went on and beat Mauler after all!

- ❍ Weight from the official scale is recorded.
- ❍ If your robot has multiple weapons, it must be reweighed in each configuration.
- ❍ MultiBot segments are weighed separately. (A MultiBot is considered disabled if 50 percent of the bot, by weight, is disabled.)

> Radio

- ❍ Frequencies are legal.
- ❍ Entrant has at least two sets of crystals.
- ❍ Any custom equipment is FCC compliant.

> Activation/Deactivation

- ❍ The robot can be completely activated in 60 seconds.
- ❍ All critical deactivation steps (such as weapon shutdown) can be performed within 30 seconds.
- ❍ All final deactivation steps can be performed within 30 seconds.

> Transmitter Turn-on/Turn-off Safety

- ❍ The robot does not move when the transmitter is turned on.
- ❍ The robot does not move when the master switch is turned on.
- ❍ The operator can reliably control motion of wheels, tracks, legs, and so on.
- ❍ When the transmitter is turned off, drive power to motion systems and weapon systems must stop immediately. (This is the test new BattleBots most often fail.)

> Internal Combustion Engines

- ❍ The fuel tank capacity doesn't exceed the maximum.
- ❍ Fuel tank is securely mounted.
- ❍ Throttle-return springs are properly designed and mounted.
- ❍ Fuel lines are properly installed.
- ❍ Fuel will not continuously spill if BattleBot is flipped upside-down.
- ❍ Engine starts in 30 seconds and goes immediately to idle speed.
- ❍ Engine returns to idle speed, or shuts off, if transmitter is turned off.

> Weapons

- ❍ All weapons must be of a type allowed by technical regulations.
- ❍ Any projectile tethers do not exceed length limits.
- ❍ Operator can control all weapons.
- ❍ Deactivated weapons pose no threat to people near the BattleBot.

> Pit Area Rules

- ❍ No children under 8 years old.
- ❍ Children under 12 must be with adults.
- ❍ No pets are allowed.
- ❍ No smoking, drinking, drugs, or unruly behavior.
- ❍ No scooters, bicycles, or skates.
- ❍ All BattleBots must be supported for runaway protection.
- ❍ All safety covers and restraints must be installed.
- ❍ No pressurized air or CO_2 refill tanks.
- ❍ No flammable liquids.
- ❍ BattleBot pressure tanks must not be left loose.
- ❍ Pneumatic/hydraulic systems must be unpressurized.
- ❍ No welding or grinding outside the designated area.
- ❍ No fueling of BattleBots outside the designated area.
- ❍ No BattleBot testing outside the designated area.

3 > THE COMPETITORS

I know it's a cliché, but BattleBot builders really do come from all different walks of life. Some you might expect: there are mechanical engineers, computer geeks, special effects artist, rocket scientists, and machinists. But wander the pits at a tournament and you'll also meet retired fishermen, high school students, lawyers, and goatherders. You'll meet athletes and paraplegics. Young girls and old men.

For some builders, the BattleBot is an end in itself—a project shared by a family, a cool piece of moving mechanical sculpture, a little something the world has never seen before.

For others, the BattleBot is a means to another end. They are here to win. Or to show off. Or to hang out with other robot lovers. Or to take part in some safe, sanctioned, no-holds-barred destruction.

BattleBot builders are as unique as their creations. Every builder, every team, has a story to tell. Here are a few of them.

BattleBot builders are as unique as their creations.
BATTLEBOTS
more on the competitors

CHRISTIAN CARLBERG [TEAM COOLROBOTS]

Before Christian Carlberg could bring any of his robots into the BattleBox arena for the first time, he first had to help *build* the BattleBox arena. "Pete [Lambertson] was super busy," says Carlberg. "So Trey [Roski] asked me to build the original BattleBox hazards." The first BattleBox had two sets of saws and two sets of lifting spikes.

Overkill battles Biohazard

"I took two weeks of vacation from work to get the units done," Carlberg says. "I borrowed trucks to move the units to Long Beach. I used to come in at about 10:00 P.M. and work until morning installing the equipment." Although Carlberg volunteered his time and effort to help BattleBots get off the ground, Carlberg says, "[Trey] said that I would have 'bragging rights.' I guess this is part of the pay-off!"

Coming to California

Carlberg's background is in mechanical engineering. He worked in aerospace for about three years before moving to California in 1995 to break into the movie special effects industry. He quickly landed a job at an effects house, where he picked up valuable hands-on experience and met some of the people who would later form his robot-building team.

Carlberg also discovered robot combat in California and began competing in local tournaments. "When we first started, we weren't doing it to get on TV," says Carlberg. "There was no such thing. We were doing it to because we thought it was the coolest thing around, and we wanted to be part of it."

In 1997, Carlberg got a call from Disney. "A Disney employee, Dan Danknick, told his boss about me and these robots I was building," he says. "Disney gave me a job on the spot. I've been working for them ever since."

Coolrobots

Carlberg still contracts for Disney and other companies, but he's determined to make robot building into his full-time job. "I've started to build a team and a small business around it," he says. "We're working to pull in sponsorship money, but more importantly we are offering our services to industry. We're using our TV time to get work building industrial robots."

It may seem strange at first that a sword-wielding BattleBot would have anything to do with industrial robotics. But that's not so, says Carlberg. "The fact is that we're building robots that work reliably under very harsh conditions. That's what industry wants." It also doesn't hurt, he says, if your robots look good on TV.

"We've developed techniques for building robots quickly," says Carlberg. For Season 4.0 of Comedy Central's BattleBots show, Carlberg shared his bot-building expertise with viewers, taking them through the process of building their own robot in a series of short, easily digestible chunks. "I built a fast, fun robot, right in front of the camera, in only about five hours."

For Season 4.0, Coolrobots brought five robots to the tournament, each unique. "No two robots are alike," says Carlberg. "I think that's what appeals to people."

"We're building robots that work reliably under very harsh conditions. That's what industry wants." — Christian Carlberg

Minion

Overkill

Toe Crusher

Dreadnought

Christian Carlberg and Toe Crusher

CARLO BERTOCCHINI [TEAM BIOHAZARD]

BattleBots cofounder Greg Munson clearly remembers meeting Carlo Bertocchini. "We had La Machine and we thought we were hot s&*%. And then Carlo came out with Biohazard and cleaned our clock and brought us back to earth." Biohazard was Bertocchini's very first combat robot. It won Robot Wars' Heavyweight championship in 1996, its first year in competition, and again in 1997. But it wasn't how often the robot won that so impressed Munson, and everyone else, as much as how well it was engineered.

Biohazard fights back at Overkill

"Carlo stunned everybody with Biohazard," says Munson. "He shows up with one small toolbox and a battery charger. He never opened the toolbox. After a fight, all he does is go back to his table, charge the batteries, get in line for the next match, win... Meanwhile, everyone's tearing their bots apart, grinding, welding, re-soldering, swapping out motors, sweat dripping down their face, freaking out that they're not going to make it to the next match in time. Carlo's over there chatting with someone. He showed everyone how it's done."

Meticulous Detail

Bertocchini doesn't have a college degree. He learned his engineering skills at Raychem, part of Tyco Electronics, where he worked as a mechanical designer, using PTC's Pro/ENGINEER CAD software to design plastic parts, injection molds, and production equipment and machines. "It wasn't too far a stretch for me to go to robot design," he says.

Bertocchini's robots are designed in meticulous detail, in software. "I like to see the whole thing together before I start cutting metal," he says. "Every part is modeled on the computer, right down to the switches and the boxes for the electronics."

Biohazard has continued to win for an amazing six years. "It's proven to be a very good design," Bertocchini says. But Biohazard is no longer able to make it through a tournament without taking any damage. "It's been an arms race with the weapons," says Bertocchini. "It's no longer the case that I can just bring it back to the pits and plug it in. The last couple of competitions I've been definitely working away on it. It's been getting damaged."

Bertocchini has had a new Super Heavyweight robot in the works for some time now, but he's a patient man and wants to get everything just right the first time. "The design is done," says Bertocchini. "They're still building the parts."

In addition to designing, building, and repairing robots, Bertocchini runs a Web site, www.robotbooks.com, which sells robot books, robot kits, and robot parts, including the Magmotor Bertocchini uses to drive Biohazard. "I worked with the Magmotor company to develop those motors for BattleBots use," he says.

The Cat's Okay

Bertocchini is known by many Comedy Central viewers as the guy who built an electric wheelchair for his crippled cat. April fools! The TV segment featuring the wheelchair-bound feline was fake; Carlo's cat is just fine. "Comedy Central wanted to do something funny," says Bertocchini, "and I came up with the Catbot for them. I intended it to come across as an obvious joke, but they did such a good job of editing it together that a lot of people thought it was real."

Tazbot combats Biohazard

"Carlo stunned everybody with Biohazard."
— Greg Munson

(left to right) **David Andres, Carlo Bertocchini, and Biohazard**

Biohazard

JIM SMENTOWSKI [TEAM NIGHTMARE]

Jim Smentowski is BattleBots' unofficial historian. Looking for the name of an early PunchBot? Or wondering who won the Super Heavyweight Rumble in May of last year? Or when Mauler 51-50 replaced Mauler? Jim knows. His Web site, www.robotcombat.com, is one of the best sources of information on the Web for all things BattleBots.

Nightmare against BattleRat

Smentowski is also one of BattleBot's most feared competitors, the man who pioneered the upward-rotating vertical spinner. The way his robot, Nightmare, tore competitors into tiny bits and tossed them through the air was an inspiration to aspiring BattleBot builders and lovers of destruction everywhere. Nightmare and its little brother Backlash have been copied by many but bested by none.

Go for It

Interested in building a vertical spinner (or less destructive robot) of your own, but don't know where to start? "Go to an event. Get a close-up view of the robots. Talk to the builders. Get on the Internet. Check out builders' Web sites. Go to the library. Read everything," says Smentowski. "One of the biggest mistakes people make is thinking they're not capable of building a BattleBot of their own. Don't sell yourself short. Get out there and give it all you've got. You may surprise yourself and a lot of other people along the way."

And don't let your limited budget stop you. "I've seen $500 robots beat up $12,000 robots. Maybe during a battle a wire will pop loose in that $12,000 machine; then it's just a really expensive chew-toy for the opponent. You're at the mercy of luck and the arena weapons."

SFX

Like many prominent BattleBot builders, Smentowski has a background in the special effects industry, but Smentowski's experience is in the software, rather than hardware, side of the business. "I worked at Industrial Light + Magic for five years," says Smentowski. "I started off as a computer technician, but I had a background in graphic design, so I wanted to get back into the creative end. I took some training and got transferred to the art department, where I did 3-D animatics and concept art using form•Z."

Smentowski still uses form•Z for the "little bit" of design work he does on the computer. He left ILM in 2001 to start his own video and Web development company. "I'm working for myself now," says Smentowski. "I wanted to free up my time, so I can choose when to work on my robots."

The Swarm

Smentowski's newest project is the Super Heavyweight MultiBot, The Swarm. The Swarm is a three-part MultiBot—essentially three Middleweight robots competing as a Super Heavyweight. Smentowski, Stephen Felk, and Paul Mathus each build one individually controlled component of The Swarm: Locust, Mosquito, and Dragonfly. Sure, it's three against one, says Smentowski. "But we're at a major disadvantage of weight. Each of us is only around 108 pounds, and we're attacking these guys that are 340 pounds. It's very difficult to inflict much damage to a bot of that size."

While none of the robots has an active weapon, Smentowski is rumored to be upgrading Locust to have more of a "bite," with one of his trademark vertical spinners.

Smentowski is one of BattleBot's most feared competitors, the man who pioneered the upward-rotating vertical spinner.

Nightmare

Backlash

The Swarm

Jim Smentowski and Nightmare

MARK SETRAKIAN [TEAM SINISTER]

BattleBots cofounder Greg Munson remembers when longtime friend Mark Setrakian was the most creative kid in high school. "He was widely recognized as a prodigy of anything he put his hand to," says Munson. "Mark's sculptures were always on display in the library. He would make these disgusting ceramic creatures—monsters with horrible things spouting out of their bodies—but they were so undeniably great that the school put them on display anyway."

Snake creeps across the arena floor

Special Effects

After high school, Setrakian attended UCLA, but he grew frustrated with the university's film program. So in 1985 he joined George Lucas's Industrial Light + Magic to start working in films for real. His first job? "I designed and built beaks for *Howard the Duck*," he says. "You've got to start somewhere."

"Up until that time," says Setrakian, "I had worked very hard at sculpting, and painting, and music ... at becoming an artist. But mechanical design came to me fairly easily, and ultimately that's what I came to do." Setrakian still works in the motion picture industry, currently at the shop of creature effects legend Rick Baker, where he's just finished work on *Men in Black 2*. "Puppeteering is also a big part of my job," he says. "I usually puppeteer the characters I design and build."

Mechadon and the Snake

Setrakian is also a veteran of robot combat from way back. He competed at every Robot Wars between 1994 and 1997, with his spherical-wheeled creation, The Master, and his serpentine robot, Snake, which moved by writhing around on the arena floor. The Master eventually became Heavyweight champion.

But Setrakian created his most famous robot, Mechadon, a 475-pound mechanical spider-like Super Heavyweight, for the first BattleBots event in 1999. "Mechadon was built specifically for BattleBots," says Setrakian. "I wanted to differentiate my BattleBots work from my work for Robot Wars." Mechadon was, he says, his love letter to BattleBots—the coolest thing he could build.

Mechadon's unique design reflects Setrakian's own conception of a fighting robot. "I didn't want to build something that looked like a lawnmower," he says. And it certainly doesn't. "Mechadon does not usually win," Setrakian says, "but it leaves a big impression. People remember Mechadon."

"I learned so much building that robot," says Setrakian. "If you build any robot—even a Wedge—you will learn a *lot* about every aspect of engineering, electronics, design, and on and on."

Designs like Snake and Mechadon are remarkable for their animalistic design and movement. "My approach to mechanical design takes nature into consideration," he says. "I view it as a useful guide, but I try to throw in a few interesting little twists."

"I'm proud of my work in BattleBots, and I'm proud to be a part of this creative robot building community."

"I'm proud of my work in BattleBots, and I'm proud to be a part of this creative robot building community."
— Mark Setrakian

Mark and The Master

Mechadon stalks

Mechadon attacks

Snake

Mark Setrakian and Mechadon

DONALD HUTSON [TEAM MUTANT ROBOTS]

Diesector meets Dawn of Destruction

Donald Hutson is the kind of guy who isn't really happy unless he's making stuff. He's worked in the awning industry, where he learned how to weld, sew, and fabricate out of metal. He spent nine years repairing medical equipment. And he's currently employed at the Neurosciences Institute in La Jolla, California, where he designs and builds custom mechanisms for scientists trying to unlock the mysteries of the human brain. Those mechanisms can range from automated camera gear to sleep deprivation chambers for fruit flies (better not to ask) to artificial rat whiskers for a robot called Nomad, which learns the way a baby learns.

The Glider Cam

But his jobs don't suffice to keep his hands and mind busy, and in 1996 his need for cool high-tech hobbies inevitably led him to robotic combat. Back then, Hutson was investing all his energy in his Glider Cam project: "I was trying to put myself inside a radio-controlled glider using a virtual reality helmet. I'm still kind of inspired by the whole idea." Hutson bought a brand-new $500 camcorder, took it apart, and reassembled it inside an RC glider. The camera sat in the glider's nose on a pan-and-tilt platform, driven by a copilot. Footage was recorded to tape onboard the glider and also transmitted to the ground.

So what happened? "I kind of dropped the project," he says, "when the glider crashed." You can ride along on the Glider Cam's plunge of death on Hutson's Web site.

Tazbot

Fortuitously, Hutson had recently heard about the sport of robot combat. In sudden need of a new hobby, he designed and built his first combat robot, Tazzbot, the first version of his current Heavyweight BattleBot, also named Tazbot.

From the start, Hutson has wanted his robots to be more agile than the others and to have more options. "I want to use my agility against my opponent's," he says. "Why put a big weapon on the robot, when I have the biggest weapons in the arena? If I can make my robot more agile than the other robot, I can use that. The arena is my friend."

Tazbot, for example, has cambered wheels for faster turning and 360 degrees of attack. In addition to biting jaws, Hutson's invertible Super Heavyweight, Diesector, wields sledgehammers that can attack at any corner, upside down or right side up. But his latest robot, Root Canal, kicks the idea of agility in the arena up another notch.

Root Canal

Hutson conceived of Root Canal while playing the computer game Quake: "I realized that I couldn't play the game without strafing [moving from side to side]. I wanted to build a strafing robot." Such a robot would be able to dodge from side to side without turning, and "circle strafe"—circle entirely around an opponent while always facing toward him. The result was Root Canal.

The key to Root Canal's mobility, says Hutson, is its omnidirectional wheels. Rather than tires, Root Canal's wheels feature six green barrel-shaped "beads," rotating freely in the hub. The motors and wheels are set radially every 90 degrees around the robot's flexible frame, at 2:00, 4:00, 8:00, and 10:00. Just looking at a still picture, it's hard to understand how the whole thing works together. You've got to see the robot in action. "There's a bunch of geometry behind it," says Hutson.

Root Canal can move forward and backward, and it turns like a tank. But thanks to special mixing algorithms in the robot's onboard controllers, and a third joystick on Hutson's end, Root Canal can also strafe, circle an opponent, or move diagonally.

"It's very fun to drive," says Hutson. "It's so agile. With strafing, you can out-maneuver anything. The weapon doesn't need to be so powerful, because it's like a little bee; it just keeps coming at you."

"Everything on every one of my robots is there for a purpose."
—Donald Hutson

Looking Good

Hutson's robots are invariably unique, and he's done well in the increasingly commercial sport. "There are two battles going on now," he says. "On the one hand, people want their robot to win, but on the other hand they are seriously concerned about it looking good."

Despite the temptations to make a more commercial-looking robot, Hutson insists that his designs are pure. "Everything on every one of my robots is there for purpose."

Tazbot

Diesector

Root Canal

(left to right) **Mike Moore, David Cook, Dawn Bernstein, Donald Hutson, David Bleakley, and Diesector**

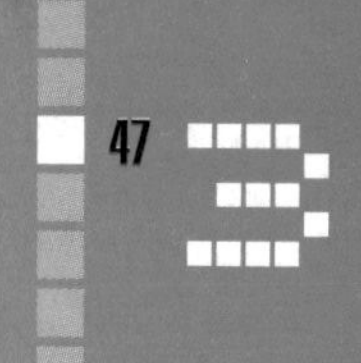

IAN WATTS [TEAM BIG BROTHER]

Robot builder Ian Watts was a broadcast engineer for the BBC for 16 years. He also lectured at Brighton University where he wrote the broadcast engineering course. So where did he pick up his robot building skills? "I've always tinkered with mechanical things," he says. "I learned to weld from building my own cars. I was too poor to buy a car. I build my first car from a wreck... took it off a scrap heap and welded it back together. I've built many cars. The current one—which has gone on hold since I'm building robots—is a replica of a Lamborghini Countach. It's three-quarters finished and being neglected." This is not an original Countach, Watts insists: "It's better than that. Lamborghini's very good at making engines, but not very good at making cars."

U.K. BattleBots, Little Sister and KillerHurtz

Build One, Daddy!

The Watts were in the Middle East for four years and missed the start of the robot combat revolution, says Watts. "We came back to England and Joe my son saw the BBC Robot Wars on the TV and said, 'We've got to build one, Daddy!' I said, 'No, no, no. That would cost lots of money and take lots of time and lots of effort.'"

But then, on a family camping trip, the inevitable happened. "We were sitting around," says Watts, "just idly chatting and throwing ideas back and forth. I got an envelope out and started sketching. That was fatal, really."

Big Brother

Watts's first robot was named Big Brother by his son, Joe. "I made everything on that robot myself," says Watts, "the motor controllers, the transmission, everything. It wasn't very good." His follow-up robot, Bigger Brother, was better. Big Brother and Bigger Brother competed in the BBC's Robot Wars. But he wanted to bring a new robot to BattleBots in the United States. "That's how Little Sister was born."

Watts does all of the design and construction on the robots and his nine-year-old son, Joe, is the weapon operator during matches. "Joe is very good," Watts says. "John Reid of Killerhurtz was driving Little Sister's arm for me on [a show we did for] Nickelodeon. John was rubbish. Missed everything."

The team's two little sisters, Megan and Ellie, aged five and six, have an important job as well. "They're the part of the team that's photogenic," says Watts.

Watts is retired from the BBC now and building robots full-time. "I still lecture at the university a bit, though I try not to," he says.

aBy

Watts is currently building a new robot, but this one's cute and cuddly. "It's for a BBC TV show called Holby City," he says. "It's a soap set in a hospital. Every couple of weeks a baby gets born, or a bit falls off and they have to go into hospital."

Another one bites the dust

The show was using silicon babies, but there was something missing from them. "A friend of mine was talking to the make-up girl on the show," says Watts. "She said, 'If only they could *move*.' And my friend said, 'Talk to Ian. He's on this robot building lark, and he could do that for you easy. And she phoned me up and said, 'Would you like to make babies with me?'"

The result is called aBy—animated baby—a silicon baby sculpture stuffed with Watts's robotic gear. aBy moves, breathes, and turns her head. "She's frighteningly lifelike," say Watts. "I just hope none of the neighbors looks in the window when I'm picking her up by her leg."

"I got an envelope out and started sketching. That was fatal, really."
— Ian Watts

Little Sister

Bigger Brother

(top row, left to right) **Joe Watts, Catherine Watts, and Ian Watts;** *(bottom row, left to right)* **Ellie Watts, Megan Watts, and Little Sister**

ZACH BIEBER [TEAM DIABLO]

Zach Bieber, creator of El Diablo and El Diablo Grande, is an artist with an eight-inch goatee and an admitted predilection for fangs, claws, and horns. Bieber grew up in a small town in Colorado. "My parents had trouble understanding me to begin with." he says. "My choice of subject matter on top of it—crazy, mangled creatures, aliens and monsters, movie stuff—was a little much for them to deal with." Bieber's mom used to refer to his artwork as his "Satan drawings"—but in a proud way. "She would have me display them for houseguests," remembers Bieber, "much to their surprise!"

As a boy, he spent his time tearing apart his mechanical toys, and much of his parents' expensive audio and video equipment, just to see what made them work. He eventually earned a degree in mechanical engineering from Colorado State University, where he built several autonomous rolling and walking robots, before heading to Los Angeles to leverage his varied skills—drawing, painting, airbrushing, sculpting, electronics, and mechanical engineering—into a career in special effects.

El Diablo meets Darkness

Too Much Fun

Bieber did work on several commercials and worked for a while creating prototypes for toys. A few were licensed to Mattel but never actually made it to the shelves. "They kept telling me that our stuff was too complex and couldn't be mass produced," Bieber says. "It was too fun, in other words.

"I hated cutting corners and dumbing down my ideas. So I went a different direction." That direction was BattleBots. The sport gives Bieber the freedom he craves. "If I'm going to do this, I'm not going to copy anybody's design. I'm not really interested in taking home a trophy, though that'd be nice. I want to come out and show a new design Put some flair into it."

Slam Job and El Diablo Grande duke it out

Thanks Heather

Bieber's girlfriend and teammate, Heather Kyseth, is integral to his design process, he says. "She offers objectivity. She has walked into the shop on several occasions and pointed out a very simple solution to a specific design that had been perplexing me for hours.

"She also reminds me to eat and sleep when I have not done either one for life-threatening amounts of time."

Bieber's mom used to refer to his artwork as his "Satan drawings"—but in a proud way.

El Diablo Grande

El Diablo

Devil Dog.
Bieber's painting based on original photo by William Wegman

(left to right) **Robert DeVaney, Heather Kyseth, Zach Bieber, and El Diablo Grande**

ROBOT ACTION LEAGUE

"Years ago," says Mike Winter, "Will Wright and his daughter Cassidy, and Jim Sellers, and my daughter Lisa and I formed the Robot Action League." The Robot Action League is not a team, per se. Rather, it is a society dedicated to saving mankind from alien robots. What? You've never been accosted by alien robots? See how well they're doing their job?

Slap Em Silly vs. Afterthought

The Winters

"I've been making robots since high school," says Winter. "My younger brothers and I would have contests, not unlike BattleBots, but on a far, far smaller scale.

"I might make a robot wander around the house, or pretend to eat cat food, or make a robot platform for my cameras. But robots weren't really that much fun at the time. Then, in 1994, I read an article in *Wired* about Marc Thorpe and Robot Wars. That sounded like the most fantastic thing ever! I phoned Marc up and he and I became friends. I came out to Robot Wars in 1994, and we've been doing it ever since."

"The best part," he says, "is hanging out in the street at 3:00 A.M. with 20 crazed robot people. You've been some wacko robot builder, all by yourself. When you get together with everybody else, with other robot people who appreciate what you do, it sort of validates your life. And it's a lot more fun."

Professionally Winter is a software designer. He was co-creator of early CAD and illustration programs for the PC, and some... other projects. "A friend and I made the game Tribal Rage, which was all bikers and lesbians beating up on each other."

Winter's daughter Lisa, now 15, has been actively involved in BattleBots since the beginning, and has gained some notoriety with her BattleBot, Tentamoushi. "She loves the combat," he says.

Winter is currently busy with the "Stupid Fun Club." A multimedia extravaganza targeted at 15-year-olds, the Stupid Fun Club combines puppets, remote control robots, and berserk 2-D animation. Winter is producing the show with fellow Robot Action Leaguer Will Wright.

The Wrights

"I met Mike [Winter] at the first Robot Wars, in 1994" says Will Wright. "We hit it off. Mike and I were on the same wavelength. It turned out we both had daughters who were about the same age. We got [our daughters] into the robot thing very early. They were the only kids involved in this for the first few years."

Like Winter, Wright is a software designer, the creator of the mega-popular games SimCity and The Sims. "That's my day job," says Wright. "It pays for the robots. The reason I work on robots is to get away from the computer."

"I've built robots as a hobby since I was a teenager," says Wright. "When the robot fighting stuff came around I really had to try it."

Wright competed at the first Robot Wars in 1994 and his daughter, Cassidy, entered her first robot the next year. Over the years, the Wrights have built a good dozen combat robots, with whimsical psych-out names such as Julie-Bot, Kitty Puff-Puff, Bob Smith, and ChiaBot.

Stretching the Rules

Both Will and Cassidy are intrigued by MultiBots—robots that split into smaller parts. Cassidy's current BattleBot, Super ChiaBot, is a MultiBot, as was the Robot Action Combat Cluster (RACC), which Will Wright built with Mike Winter. "I had half of RACC," says Wright, "and Mike had the other half. His was a spinner, and mine was a thwacker, but mine also had a third robot that would pop out of it. We called it the emergency escape pod, or eep. Eep was this tiny little robot made out of foam and servos. I designed it to be indestructible. The robot was way down inside the foam. No matter what you hit it with, it would just bounce away."

At the time, a MultiBot was considered disabled if 50 percent of the components were disabled. If one of the

"The best part is hanging out in the street at 3:00 a.m. with 20 crazed robot people."
— Mike Winter

big robots died, Wright would launch the escape pod, and RACC would be a two-part MultiBot again. Current BattleBots rules define MultiBots by weight, rather than number, rendering the eep strategy ineffective.

"In Robot Wars," says Wright, "I had a robot that dispensed tape. I found this really gnarly tape, and I would drive around the other robots in circles and wrap them up in it." At some point, tape was outlawed in robotic combat. "I've been trying to see how many rules I can get in the rulebook," says Wright.

Jim Sellers

The league's fifth BattleBot veteran is Jim Sellers. "Jim is the most fantastic machinist," says Mike Winter. "And he's a very creative, wacky person."

Seller's background is a little more common for a bot-battler. "I've always been a tinkerer," says Sellers. "I studied biology, biochemistry, and electrical engineering in school and finally graduated with a degree in biology. I know, odd mix of studies. I've worked as a production engineer, a design engineer, a electromechanical technician, a machinist, a welder, a furniture maker, and probably a few more things over the past 30 years. It takes a varied background to build robots!"

"Basically," says Sellers, "I'm a team of one." But that's okay. Because he's also a member in good standing of the Robot Action League!

Afterthought

ChiaBot

Tentoumushi

(left to right) **Jim Sellers, Mike Winter, Lisa Winter, Will Wright, Cassidy Wright, and their fleet of BattleBots**

NOLA GARCIA [TEAM LOKI]

"Team Loki is like a family," says team matriarch Nola Garcia. "Team Loki is mostly my son's friends. They've all grown up around my house, and they were friends all through college. Everybody hangs out together. We all get together and watch BattleBots every week."

Team Loki labs

Team Loki traces its origins back to 1996, says Garcia: "A bunch of the guys were in high school and were part of a FIRST [robot] team. I wound up being the team coordinator.

"As the kids graduated from high school, a lot of them stayed here and went to Florida International University College of Engineering. They were mentoring high school kids [in robotics], but they wanted to do something on their own. Korey [Kline] and I saw BattleBots and we said, 'This is it.'

"Korey Kline is the Founding Father, the real mastermind behind Team Loki," insists Garcia. "He is the head engineer. He is a *real* rocket scientist in the *Guinness Book of World Records*."

Team Loki/Team Fembot

The team first competed in BattleBots Season 1.0, but under their high school FIRST team name of Rammtech 59. "They wanted to have a different team name, just for BattleBots," says Garcia. "We had a big team vote and they chose Loki. Loki is the Norse god of mischief. I can't tell you how appropriate that is." Part of Team Loki's spice, says Garcia, comes from its multinational flavor; Team Loki boasts members from Ecuador, Peru, Cuba, and the United States.

Team Loki is currently fielding eight robots, including Rammstein, Turbo, Surgeon General, and Buddy Lee Stay in Your Seat, under three different team names. "Team Loki is kind of the umbrella," says Garcia. "Under Team Loki we have Team Loki 1, and Team Loki 2, and Team Fembot."

Team Fembot—Nola Garcia and Mercy Rueda—was the first all-female team in BattleBots. The team and its robot, Buddy Lee Stay in Your Seat, came about because of an incident at Robot Wars in London. "The guys wouldn't let me drive Rammstein," says Garcia. "I got a little miffed and I said, 'Fine! I'll build my own robot.'" Buddy Lee was designed on the way home, in the air between Brussels and Newark. Fembot teammate Rueda was recruited at Florida International University, where Garcia worked: "I begged her, 'Please! There's too much testosterone on our team. I need another girl.'"

You Know, for Kids

Garcia and Rueda are happy to set an example for other women out there, but, for Garcia at least, robot combat is mostly about kids. "My whole thrust in life is kids," she says. "I am shameless when it comes to begging for kids." Garcia oversees the BattleBots IQ program for high-school-aged kids, and she and her husband, Bill Garcia, have started a new facility for kids in Miami, called Starbot. Team Loki mastermind, Korey Kline, is also very involved with Starbot, says Garcia.

"We call Starbot a playground for the mind," Garcia says. "It's two 1200-square-foot warehouses with a full machine shop and a bank of computers. It's a place where kids from nine different high schools and two different colleges can come and build science projects and robots. We can't kick kids out of there.

Abaddon vs. Rammstein

"About 50 percent of the kids are girls. We have kids from the very, very poorest parts of Miami as well as kids from the very, very richest part. I guarantee you can't tell me which kids come from which part of town."

Garcia and Rueda are happy to set an example for other women out there, but, for Garcia at least, robot combat is mostly about kids.

Buddy Lee Stay in Your Seat

Rammstein

Turbo

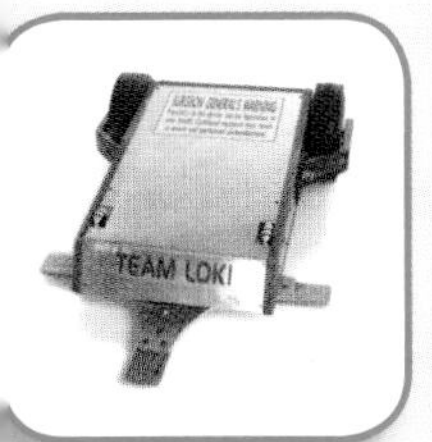

Surgeon General

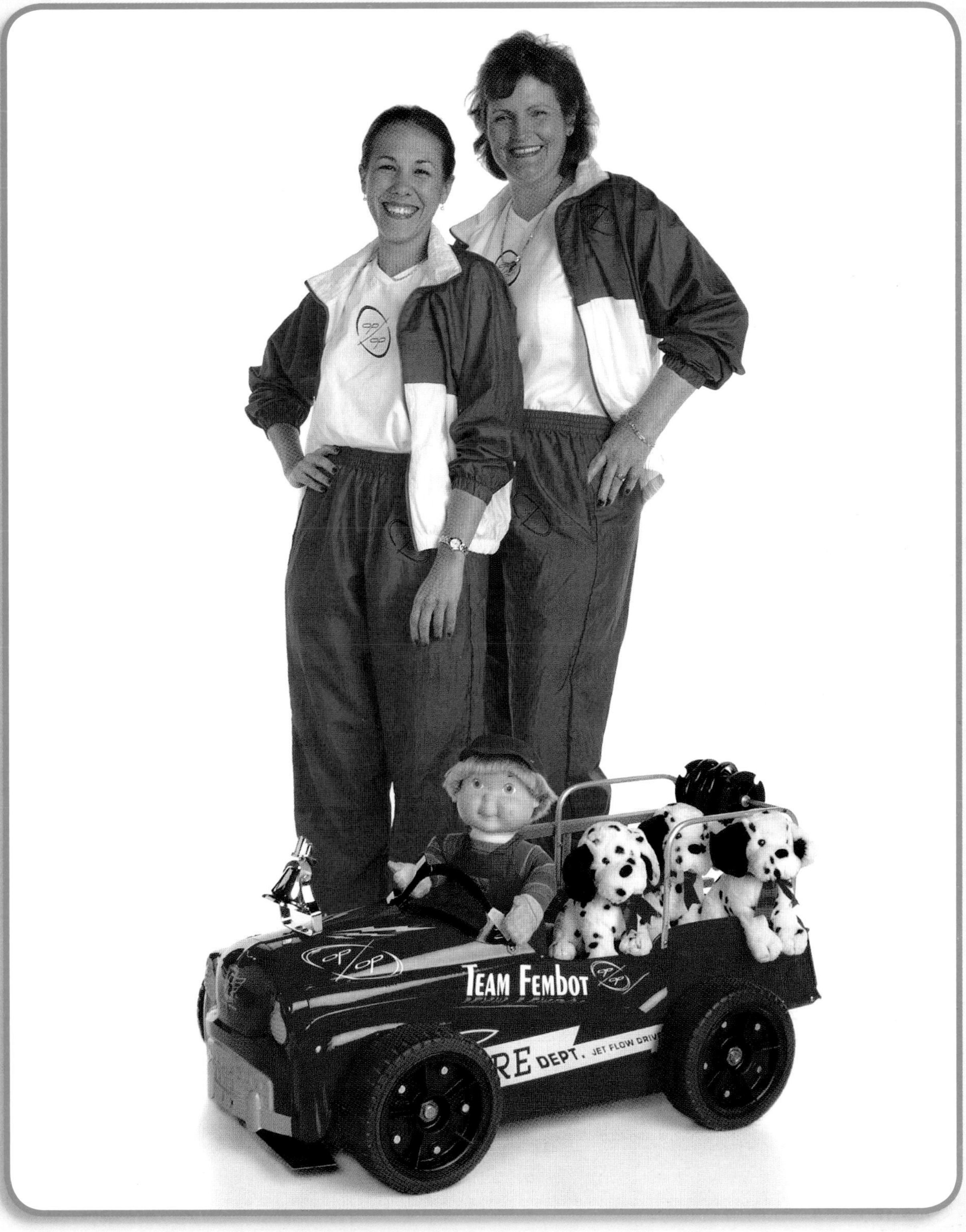

(left to right) **Mercy Rueda, Nola Garcia, and Buddy Lee Stay in Your Seat**

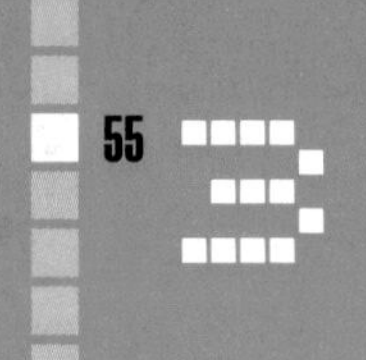

STEVE AND LOWELL NELSON [TEAM S.L.A.M., TEAM K.I.S.S.]

Steve Nelson and his dad, Lowell, have always done manly stuff together. They worked together as commercial fishermen for 17 years. They competed in tractor pulls from 1980 to 1985, winning the California State Championship twice. "Tractor pulling's kind of like drag racing at 15 MPH," says Steve. "We had a 700-horsepower Chevy engine in that tractor.

"We spent a lot of hours on it," he says. "They're almost as bad as robots. We worked on it in the carport, because the garage was full of junk. We'd fire it up at about 3:00 A.M., checking for oil leaks, and the sheriff would come by. You could hear it a mile away."

Father and Son, Together Again

But after Lowell Nelson retired and son Steve moved out of state, the two hadn't worked on anything together for years—until they discovered robot combat. "It's got us doing things together again," says Steve. "Dad and I went to Robot Wars in 1997, and while we were watching, we said, 'Hey, we can do this!'

"We came up with a design and went out and built our first robot, S.L.A.M. We were ready for Robot Wars in 1998; we were going to go take on the world. Unfortunately, Robot Wars 1998 didn't happen."

What did happen was the infamous fight under the freeway in Novato, California, in 1998; which the Nelsons initiated by challenging Jim Smentowski to a duel. The Nelsons finally got to take part in some robot combat action, and they loved it.

From Bed Frames to BattleBots

When the first BattleBots tournament took place in Long Beach in 1999, the Nelsons were there with not one, but two robots: S.L.A.M., and a new robot named K.I.S.S."Trey Roski said they needed more robots," says Steve. "So we threw K.I.S.S. together in about nine days. It was all recycled parts. I mostly built it out of a lawn tractor." The Nelsons use lawn tractors, sewer pipe, old iron bed frames, blocks of wood... whatever materials were on hand. "My steel's just as good as your steel. I don't care where you bought it.

"When we built S.L.A.M, we took a 28-inch diameter sewer pipe, set it on the floor and said, okay, all the parts have to fit in this circle," says Steve Nelson. His current project is a upgrade to his Super Heavyweight, Electric Lunch. "With this one all the parts must go in a square box. I've got six wood blocks on the table and I'm lining things up with them until I can turn it into metal. I'm in the 'pile of parts, wood blocks, and does it fit?' stage."

To many competitors, the Nelsons and their enthusiastic, low-tech approach epitomize the heart of the sport of BattleBots: two regular guys working in their freezing garage, doing it for the love of the game.

"Yep," says Steve. "We're pretty regular. I'm going to weld something now. It's starting to get cold in here."

The Nelsons and their enthusiastic, low-tech approach epitomize the heart of the sport of BattleBots.

Electric Lunch

Half Gassed

S.L.A.M.

K.I.S.S.

(left to right) **Steve Nelson, Lowell Nelson, and Half Gassed**

ARNDT ANDERSON

Arndt Anderson, builder of Heavyweight BattleBot Shark Byte, is a machinist by trade. A high-school dropout, Anderson got his first job in a machine shop through a friend. "I started as a shop squid and worked my way up to where I am now," he says. Anderson has built parts for the space shuttle, the Stinger missile, and the Patriot anti-missile missile. He designs and builds industrial robots for automated assembly lines. But he's best known, at least outside the BattleBots community, as the guy who makes magnesium guitars.

Shark Byte and Matador

Magnesium Guitars

"I play guitar," says Anderson. "When I went to work in a metal shop I thought, hey, has anyone ever built an aluminum guitar? So I built one, and it sucked. Aluminum's a terrible metal to make a guitar neck out of—it's heavy, it doesn't have a lot of memory, it expands and contracts, it feels cold and clammy to the touch.

"But magnesium was the perfect alloy. It's got a high tensile strength to it. It wants to snap back, which is essential for keeping in tune. I made one for myself and other people said, 'That's pretty cool! Can you make me one?'"

BattleBots Is the Jack

But Anderson has put the whole guitar thing on the back burner for now. "I'm supposed to be building one for [Def Leppard guitarist] Phil Collen right now," he says. "It's sitting there three-quarters done. But I've been putting him off, because I'm working on BattleBots. BattleBots is more enjoyable, and I get more recognition for it. The rock stars took all the credit for the guitars.

"When you've got Jim [Smentowski] and Nightmare out there just obliterating something, that's what interests me, total destruction. BattleBots is the jack!"

Magnesium guitar

Anderson's known as the guy who makes magnesium guitars.

Shark Byte

Anderson on guitar

Anderson goes multi-platinum

(left to right) **Cody Rose, Arndt Anderson, and Shark Byte**

MICHAEL "FUZZY" MAULDIN [TEAM TOAD

"I've been known as 'Fuzzy' forever," says Michael Mauldin, gentleman of leisure and patriarch of Team Toad. "It's my business *nom de plume*. In college I was the only white guy with an afro. Since my hair started thinning, I've grown a beard so people don't wonder [about the name]."

A cheering Team Toad

"People confuse me with the inventor of fuzzy logic," says Mauldin. "No. That was Lotfi Zadeh, not me." The confusion is understandable. Mauldin holds a Ph.D. in artificial intelligence (AI) and was, until a few years ago, on the research faculty at Carnegie Mellon University, doing research on information retrieval and machine translation.

The Chatterbot and the Search Engine

"Depending on who you ask," says Mauldin, "the most famous thing I've done is either Lycos or Julia."

Julia is a "chatterbot," a program that converses with people and, hopefully, fools them into thinking she is real. Julia's quite the celebrity within the AI community.

Lycos is, of course, a prominent Web search engine and portal. Just a little something Mauldin thought up and built in his spare time. "I put it up on the Web and people started using it," says Mauldin. "The university and I commercialized it and sold it to CMG Interactive in Boston. I got insanely lucky."

The Lazy Toad

"The net result is that I'm retired, and I'm living my ego ideal, which is Dick Van Dyke in *Chitty-Chitty Bang-Bang*." When Mauldin needed to be back home by 9:00 A.M. the morning after his first competition, he chartered a private jet. "I highly recommend private jets," Mauldin says with a smile. "It's a great way to travel."

Mauldin spends most of his time at the Lazy Toad Ranch, his 11-acre hobby farm east of Pittsburgh. The ranch is named for Walter, the family toad, and for Mauldin as well. "Now that I'm retired," says Mauldin, "I'm very lazy. So I contributed lazy and Walter contributed toad."

One for Every Kid

Why would a lazy man build a BattleBot in the first place? Mauldin has partial custody of his son Danny, and he wanted a project he and Danny could work on when they were together. What better than BattleBots? "It's a great father and son thing," says Mauldin.

"But I'm fair," he says, "so I told my daughters, 'Danny and I are going to build a robot. Do you want one too?' And of course they did, so every kid had to have a robot."

Altogether, Team Toad fields five BattleBots—Snowflake, WindChill, FrostBite, IceBerg, and Ice Cube—one for each kid, and two for Dad. And Mauldin's wife Debbie is planning on bringing her own Lightweight BattleBot, SnowCone, to the next tournament.

Danny, now nine, has only been able to go to one BattleBots competition so far. The girls, Jacey, 13, and Kelsey, 11, have had much more success with their BattleBots. "They have more driving experience in the box," says Mauldin, "and they've got better robots. I built Danny's robot WindChill first, and I made a poor motor choice. We'll be building another WindChill with better motors."

All the Time in the World

Mauldin is currently teaching himself to weld. "My wife bought me a MIG welder for Christmas," he says. "For the next season, I'm going to build a new body for IceBerg from scratch. If that works out, I'll probably [build the bodies] for all my robots in the future.

"Being retired, I can spend all my time working on these robots and learning new skills. The biggest advantage I have over my competitors is time."

"I'm living my ego ideal, which is Dick Van Dyke in Chitty-Chitty Bang-Bang."
— Michael Mauldin

Snowflake

WindChill

FrostBite

IceBerg

Ice Cube

(top row, left to right) **Debbie Mauldin, Jacey Ross, Michael Mauldin, and Joe Choo;** *(bottom row, left to right)* **Kelsey Ross, Danny Mauldin, and IceBerg**

PART TWO

THE BO

> 4— ROBOT TYPES > 5— THE ROBOTS

In only three years of tournaments, BattleBots has already seen nearly 700 unique robot competitors. Although many of these fall into the general category of "box on wheels," others defy attempts at description.

They may not come in *all* shapes and sizes, but within the allowable weight limits of 25 to 408 pounds, you'll find everything from boxes to wedges to saucer-shaped domes. Beetle-shaped bots. Spider-like bots. Beautiful bots and ugly bots. Bots with turrets and arms and spikes and chainsaws. Bots on wheels and tracks and feet. Bots that spin, hop, shuffle, and writhe on the ground. Bots that punch, and bots that hammer and chop. Bots that can lift a car. Bots that can turn somersaults in the air. Bots sporting titanium armor, and bots covered in plastic plants. Bots that cost $50,000, and bots that cost 500 bucks. Bots built by large teams in machine shops, and bots built by one guy working weekends in his garage.

Although there are strict rules and regulations governing the size, armor, weapons, and power that a BattleBot can employ, the number of possible legal robot designs is quite literally infinite. At least.

ROBOT TYPES

Mama always said there are two kinds of people in this world: those who divide everyone up into two kinds of people, and those who don't. I'm one of the former. So, as is my wont, I'm going to set out to divide all the BattleBots in the world into two kinds of bots. Wait. No, ten. Ten types of bots. Better make it a dozen.

Unlike mass-produced cars and bikes, each BattleBot is a very individual creation. Very few BattleBots could be mistaken for each other on the street. There are almost as many BattleBot types and strategies as there are BattleBot builders. *(In fact, as BattleBots evolve over time, they come to resemble their builders more and more. Spooky but true.)*

Each BattleBot, no matter how unusual in appearance, consists of the same six groups of components:

1> Motor(s) and Drive Train

Every BattleBot needs at least one motor or engine to provide locomotive power, something to push it around the BattleBox. A few BattleBots move on caterpillar treads, or walk, or shuffle, or writhe around on the ground like snakes, but by far the most common BattleBot drive train consists of electric motors coupled to wheels, either directly, or by some kind of chain or belt drive.

2> Primary Power Supply

Those motors need power—the more the better. In the case of a gasoline engine, that power comes from a tank full of fuel. For the more common electric motors, the power source is a bank or two of powerful batteries, usually either Sealed Lead Acid (SLA), Nickel Cadmium (NiCad), or Nickel Metal Hydride (NiMH). (Wet cell batteries are strictly *verboten* in a BattleBot.) Batteries are heavy and you can put them almost anywhere in your BattleBot, making battery placement a great way to tune a BattleBot's balance.

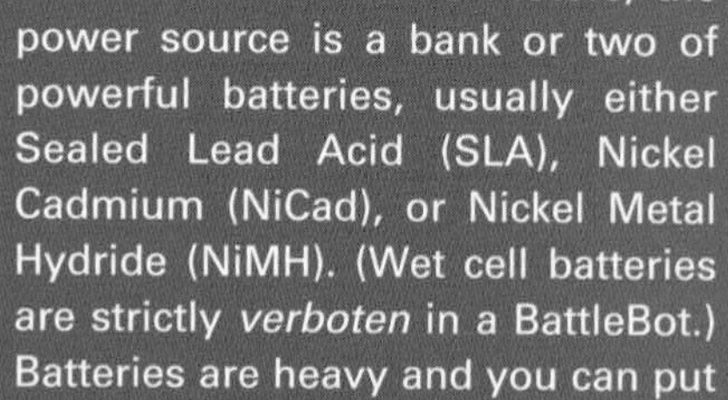

Remember: A BattleBot needs to be able to take serious hits. Armor is important, but equally important are a strong frame and internal impact-proofing. Build your drive system so it is not dependant on your overall chassis alignment, leave generous clearance around moving parts, and leave a little slack in all your wires so that connectors don't pull free if a component shifts position. Heavy components such as batteries and motors should be well secured.

3> Radio, Speed Controller(s), and Assorted Other Electronic Gizmos

Builders control their BattleBots from outside the BattleBox by remote control. Each BattleBot has a radio receiver onboard, which conveys the driver's commands to the bot's various components. Speed controllers vary the voltage going from the batteries to the electric motors, controlling the motors' speed and direction. Other electronics control weapon systems, onboard computers, and other gear. The electronics usually have their own, much smaller, power supply, separate from that of the main motors.

4> Weapons Systems

BattleBots sport a wide range of weapons, from saws to hammers to pneumatic spikes to "rotary bludgeoning devices." Some BattleBots have no real weapons at all, preferring to just shove their opponents around the BattleBox and let the hazards do the damage for them.

5> A Nice, Sturdy Frame

Every BattleBot has some kind of frame or chassis to bolt all these components to. The frame can be anything from old angle-iron to expensive, high-tech alloy, bolted or welded together, or machined in one big (expensive) piece.

6> Armor Over the Top

Strictly speaking, you could create a BattleBot without armor, but it wouldn't last very long against angry opponents bristling with spikes, saws, and hammers, not to mention the hazards of the BattleBox itself. Armor is usually tough metal-titanium, steel, or aluminum alloy—but some bots are armored with bulletproof plastic. The armor can be bolted directly to the frame or shock-mounted to absorb hits. In some BattleBots, the armor actually becomes the frame, either completely or in part. In this case, components are bolted directly to the armor plating.

You're All Set. *That's it. Put that all together (in accordance with the Official BattleBots Technical Regulations, of course) and you're done, except for getting t-shirts made up and hyping your creation on the popular BattleBot forums.*

THE WEDGE

The Wedge has the distinction of simultaneously being the most popular and least popular type of BattleBot. Watch a BattleBots match and you may see spectators waving signs that read "Wedges are for Wimps." And worse. Why? People hate Wedges because they are so simple—*and* so effective. Nobody would lavish such strong feelings on a design that never won. And yet, for some people, Wedges violate their engineering esthetic: Wedges don't seem sufficiently sophisticated to be given respect. They find them, well, boring.

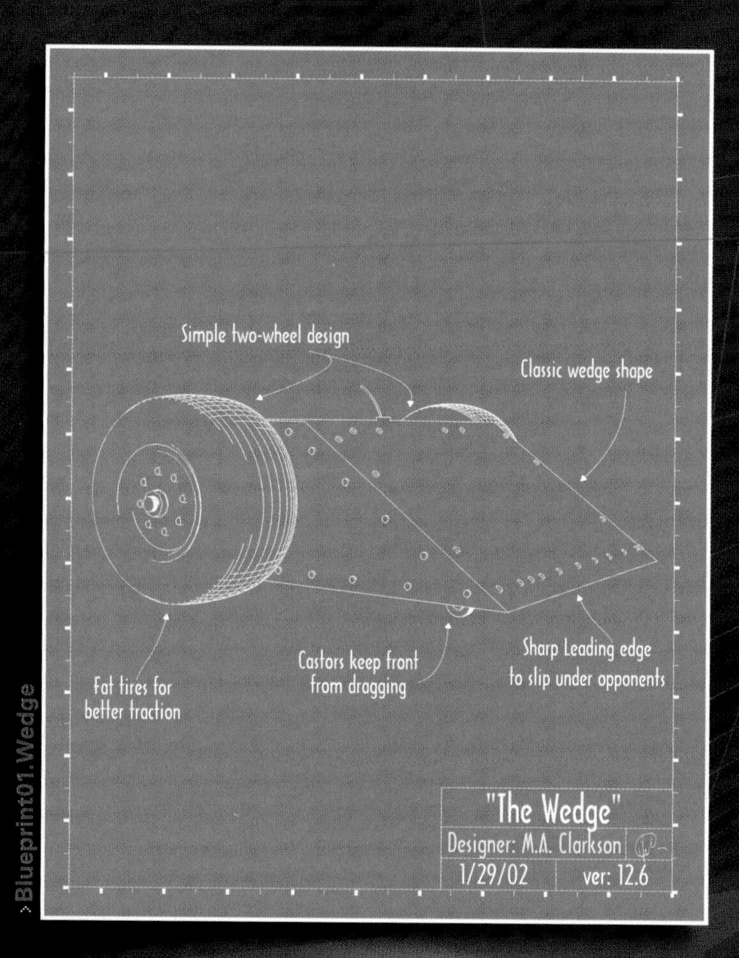

Blueprint01.Wedge

In truth, the Wedge is a wonderfully elegant design. The Wedge's weapon is its own shape, powered by its own speed and inertia. As you no doubt recall from Mrs. Richardson's seventh grade science class, a wedge is a moving inclined plane or ramp. The wedge moves farther forward than the object it's lifting moves upward, so it pushes up with a correspondingly greater force. (Work equals force times displacement.) That is, by pushing forward five inches with 20 pounds of force, a wedge can lift 30 pounds two inches into the air.

Wedges attack by driving their leading edge under an opponent's skirts—hopefully lifting its wheels (or treads or feet) off the ground—and driving opponents into the hazards.

Wedges generally have two wheels, although some have four. The leading edge of the Wedge needs to be very strong, as it will be bearing the brunt of its own attacks, as well as damage from BattleBox hazards and opponents' weapons. If it gets bent, the bot's effectiveness is greatly reduced.

> **Turn-ons:** SpinBots and RamBots. A well-armored Wedge can successfully attack SpinBots if the front of the Wedge is strong enough to survive hits from the SpinBot. A Wedge also has an edge when fighting a Ram, as the Wedge can get under the Ram, denying the Ram the traction it needs to push back or get away.

> **Turn-offs:** Lower, faster, more powerful Wedges, and ClampBots. Wedges are vulnerable to these bots, as they can lift the Wedge's wheels off the ground, rendering it helpless.

Dreadnought

People hate Wedges because they are so simple—and so effective. Nobody would lavish such strong feelings on a design that never won.

> THE RAMBOT

Nothing says brute force like a RamBot. A RamBot is conceptually even simpler than a Wedge: it's a box; it rams. RamBots often have no real weapons at all. Rather, they use their superior power and traction to shove their opponents around the BattleBox. The archetypal RamBot is just a box with wheels, although many sport some sort of blade, bumper, or "cow-catcher" attached to the front, or spikes and other scary bits.

To win, a RamBot needs more power, or at least better traction, than its opponent. A RamBot, after all, needs to push at least twice its own weight. To that end, pushers generally have four or more large, high-traction, soft rubber wheels, driven by the largest drive motors and batteries their builders can cram in. Strong RamBots have as much as 1 HP for every 10 pounds of total weight. Without weapons, they have the advantage of being able to devote all their weight to power, traction, and armor, making them especially strong, quick, and fast.

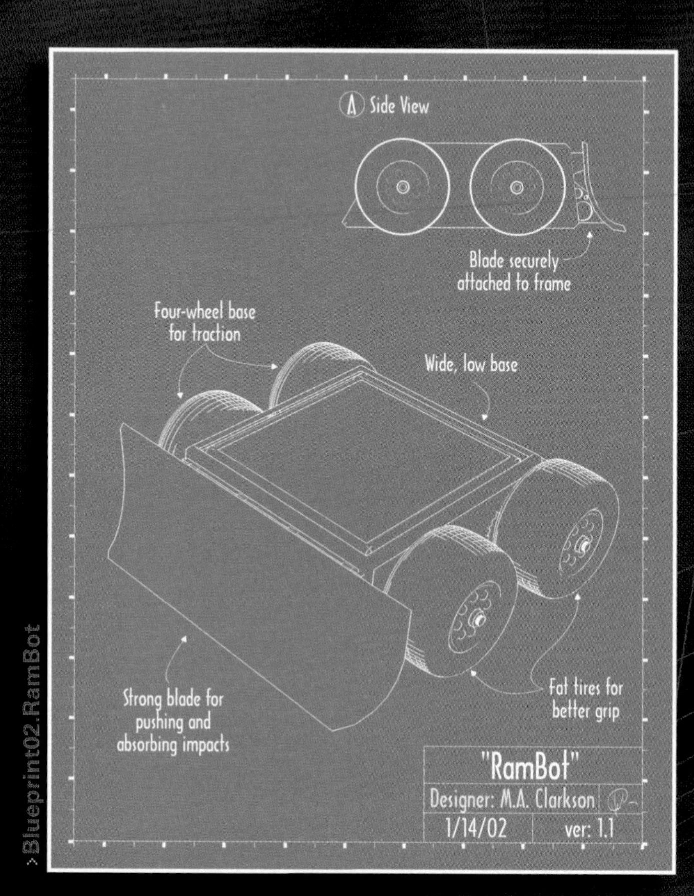

>Blueprint02.RamBot

> Turn-ons: Rotary weapons. With no fragile external mechanisms and the ability to take solid hits, a RamBot can keep hitting a SpinBot or RotaryBot until that bot self-destructs.

> Turn-offs: Wedges, LiftBots, and LaunchBots. This design is weakest against an opponent that can lift its drive wheels off the ground.

FrostBite

It's a box. It rams. 'nuff said.

THE SPINBOT

SpinBots attack by spinning their entire outer shell at tremendous speed. Everybody's got a favorite BattleBot type, and these bad boys are mine. SpinBots are the most destructive BattleBots in the arena, thanks to their effective harnessing of kinetic energy.

Kinetic energy refers to the work—or, for BattleBots, the destruction—something can accomplish because of its motion. (The kinetic energy of an object of mass *m* moving with velocity *v*, is $mv^2/2$, if you want to know.) Consider: A bullet sitting on a table has no kinetic energy and is quite safe. A bullet moving at Mach 1.5 has quite a bit of kinetic energy and is quite dangerous. The faster something is moving, the more kinetic energy it has. And the SpinBot design allows weapons to build up velocities of hundreds of miles per hour.

As the bullet example shows, an object need not be very heavy to possess considerable kinetic energy. Still, the more massive something is, the more kinetic energy it has, as well. SpinBots' weapons—bolts, padlocks, or welded pieces of tool steel—are considerably heavier than a bullet.

A SpinBot works by turning its entire outer shell into a weapon-studded flywheel—a sort of kinetic energy battery. As the weapon's motors spin it up faster and faster, the amount of energy stored becomes greater and greater. When any part of that flywheel contacts another object, like, say, a BattleBot ... BANG! All that energy is delivered in an instant, often with spectacularly destructive results.

Many BattleBots use kinetic energy weapons to do damage, but none apply as much energy, to do as much damage, as do SpinBots. But remember also that for every action, there is an equal and opposite reaction. When the SpinBot's weapon makes contact with an opposing BattleBot, not all of that energy is delivered to its opponent. The SpinBot absorbs quite a bit of that energy itself, from the kickback. (Imagine swinging an aluminum baseball bat as hard as you can at a telephone pole.) Many a SpinBot has self-destructed on impact, damaging an opponent but taking itself out of the match in the process. A SpinBot needs to be built as ruggedly as possible to avoid this fate.

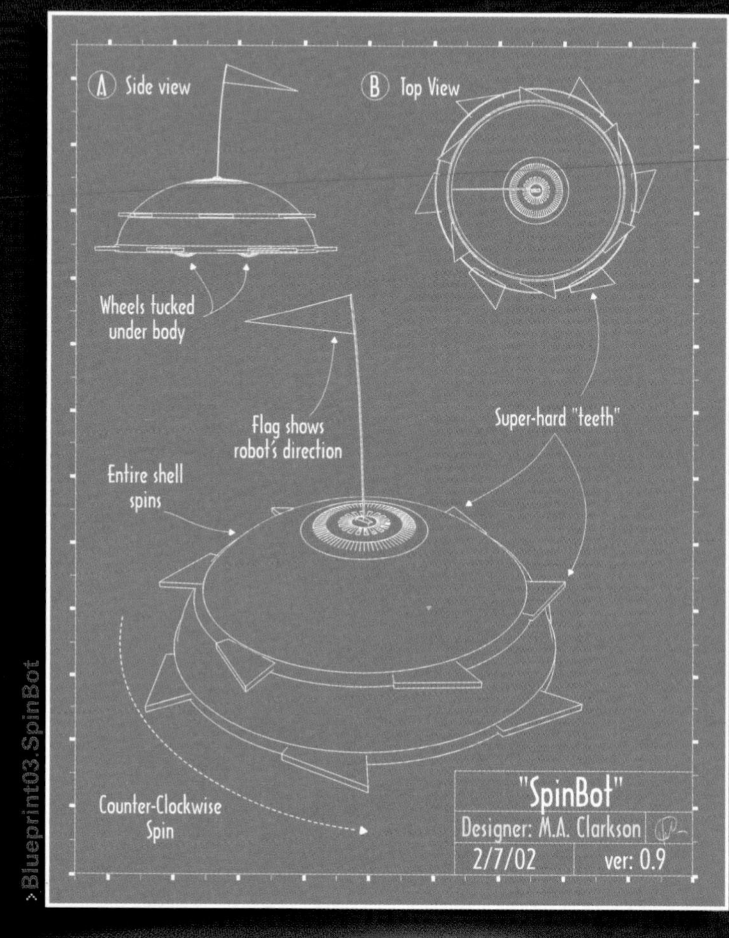

>Blueprint03.SpinBot

A fully enclosed SpinBot has an additional difficulty not faced by other bots: when the weapon is running, the driver can't tell which direction the bot is facing! SpinBots usually sport a non-rotating flag or arrow sticking up through the center of the shell, or they have a "tail", to show the driver which way they're pointing.

The spinning shell produces a strong turning force on the base of the robot, which will make the bot want to curve in the opposite direction when moving. Many SpinBot builders incorporate gyroscopes in their control electronics to compensate for the effects.

A SpinBot's offense is also its best defense. Since the entire outer shell is spinning, it's impossible for an opponent to hit the SpinBot without being struck by the SpinBot's weapon.

Turn-ons: Lifters, ClampBots, PoundBots, and ChopBots. SpinBots fare well against bots with exposed weapon parts that can be bent or broken off.

Turn-offs: Solidly built RamBots and Wedges. A SpinBot's worst possible opponent is one that can take repeated impacts until the SpinBot breaks itself. A high-speed collision with a Wedge can cause some SpinBots to flip themselves onto their backs.

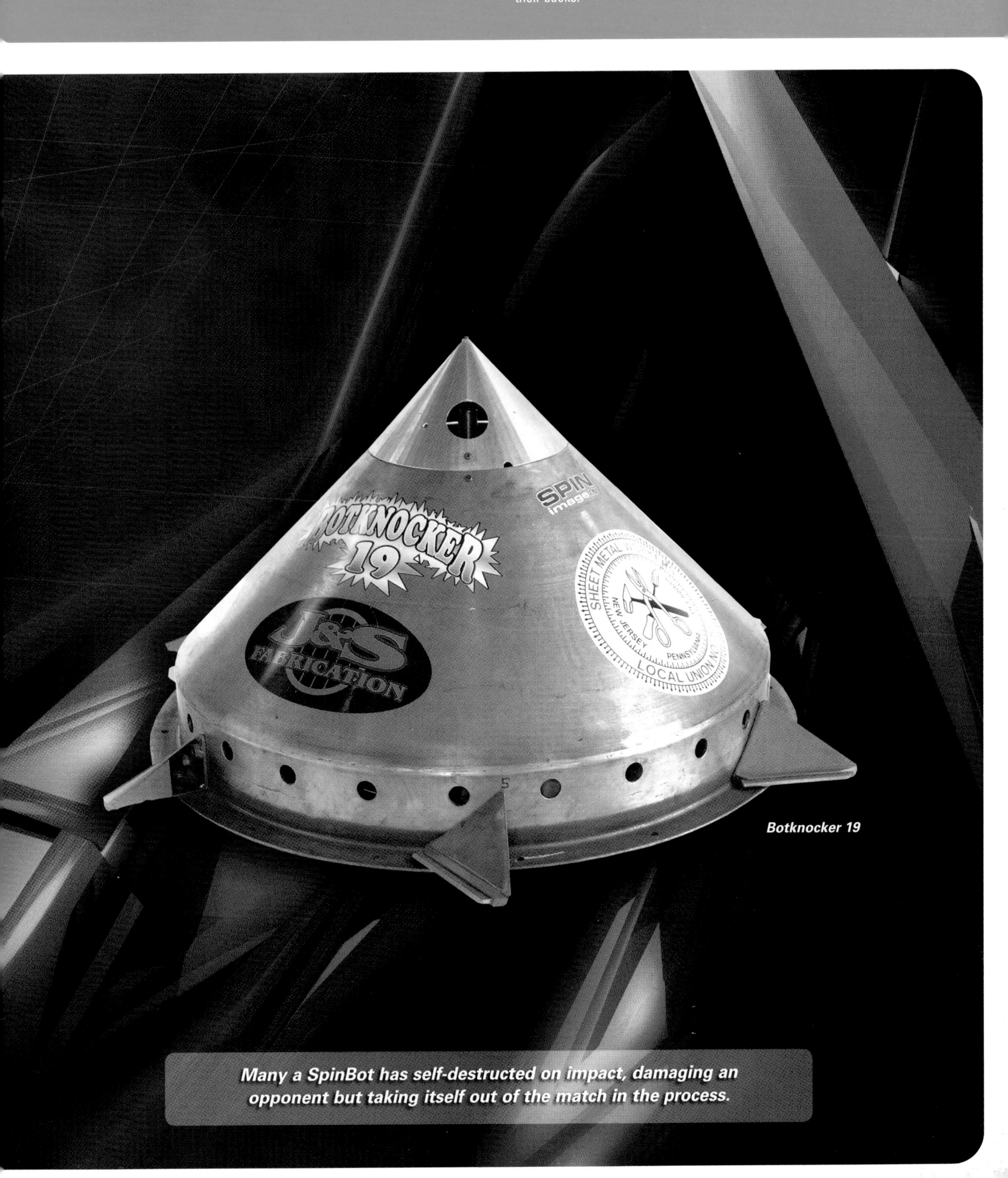

Botknocker 19

Many a SpinBot has self-destructed on impact, damaging an opponent but taking itself out of the match in the process.

> THE ROTARYBOT

The RotaryBot is a close relative of the SpinBot. Like SpinBots, Rotaries can build up lots of nasty kinetic energy and unleash it against their opponents' armored hide in a fraction of a second. But where the SpinBot spins its entire outer shell, the RotaryBot spins a blade, bar, or disc at very high rates of speed.

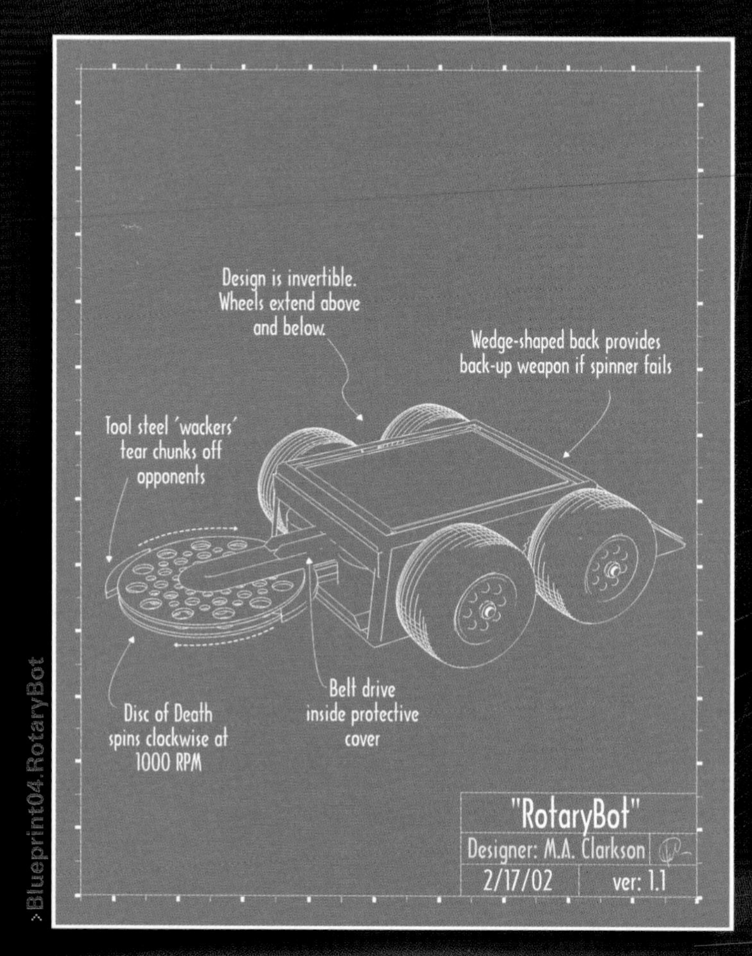

>Blueprint04.RotaryBot

The harder the teeth and the faster they're moving, the more damage they do when they hit. Consequently, builders try to drive them with the biggest gasoline engines or electric motors they can. The torque, especially from very large flywheels, can become ferocious.

Most RotaryBots spin their weapons horizontally, like a lawnmower blade. But some of the newer designs feature large, vertically spinning disks. The disks generally rotate upward, so that contact drives the opponent up off of its wheels and possibly flips it over. Oh, and tears off great big chunks.

A Vertical RotaryBot is an extremely deadly design, but also very hard to drive. A large disk gives the Vertical RotaryBot a dangerously high center of gravity that, coupled with the significant gyroscopic forces of the large, spinning disk, make it prone to tipping and difficult to turn. Vertical Rotaries require a large, wide body with widely spaced wheels for support. Despite this, the Vertical Rotary must be able to maneuver well enough to choose its angle of attack, lining the disk up carefully and protecting its vulnerable sides and rear from attack. The recoil force on the Vertical RotaryBot slams it down against the floor, and it slams it down *hard*. Vertical RotaryBots must be very well supported to avoid breaking motor mounts, axles, or weapon supports.

In fact, all RotaryBots, like Spinners, are susceptible to damage from their own kinetic energy weapons, and many a match has been lost through self-destruction.

Turn-ons: Just about anything that cannot outmaneuver them. Horizontal RotaryBots are difficult to get near, but Vertical RotaryBots are vulnerable to fast, maneuverable opponents that can stay away from their weapon and attack from the back or side.

Turn-offs: Fast, maneuverable opponents that can stay away from their weapons. Vertical RotaryBots are especially vulnerable to quick opponents that can attack from the back or side.

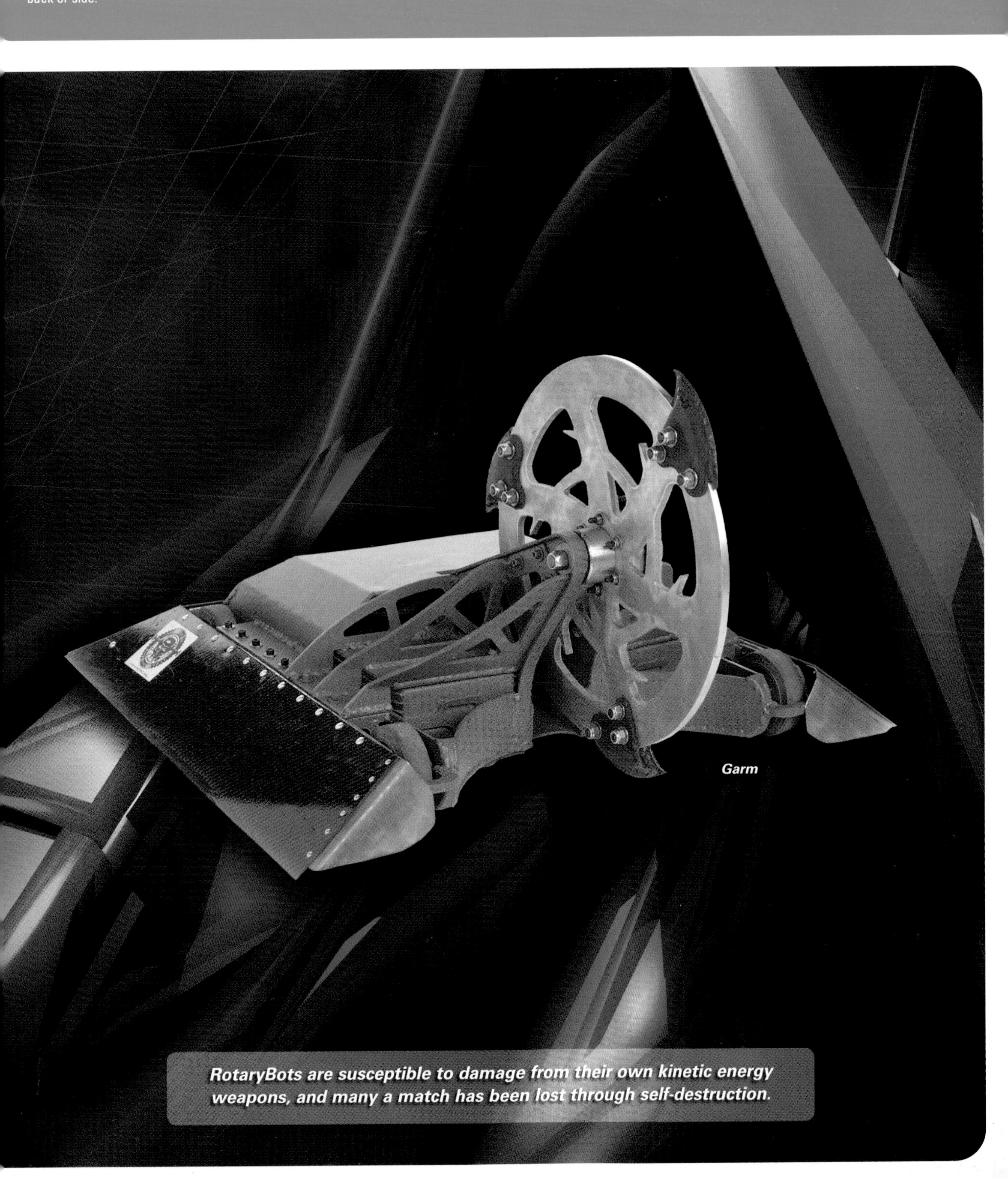

Garm

RotaryBots are susceptible to damage from their own kinetic energy weapons, and many a match has been lost through self-destruction.

THE ROTARY SAWBOT

Where RotaryBots are designed to bash, Rotary SawBots spin a disk with hard, sharp teeth or abrasive material, designed to cut and rend. Saws that spin downward on the opponent are more likely to do sustained damage, as the force of the saw's rotation helps pin the opponent down to the arena floor. Saws that spin upward from the floor, on the other hand, can also function quite nicely as launchers, since the saw's rotation tends to lift opponents off their wheels and even flip them over.

The best saw blades to use are those employed by fire and rescue workers for emergency rescue extrication. These are thick steel disks coated around the edge with hard abrasive to cut a wide variety of material quickly—just the thing for a combat situation. However, they are heavy, expensive, and available only through certain specialty dealers, and they require a seriously powerful motor to be used to full effect.

In fact, SawBots are increasingly rare. The saw, by itself, is usually not very destructive; most BattleBots today are just too damned hard to cut through, especially when they're trying to run away from you.

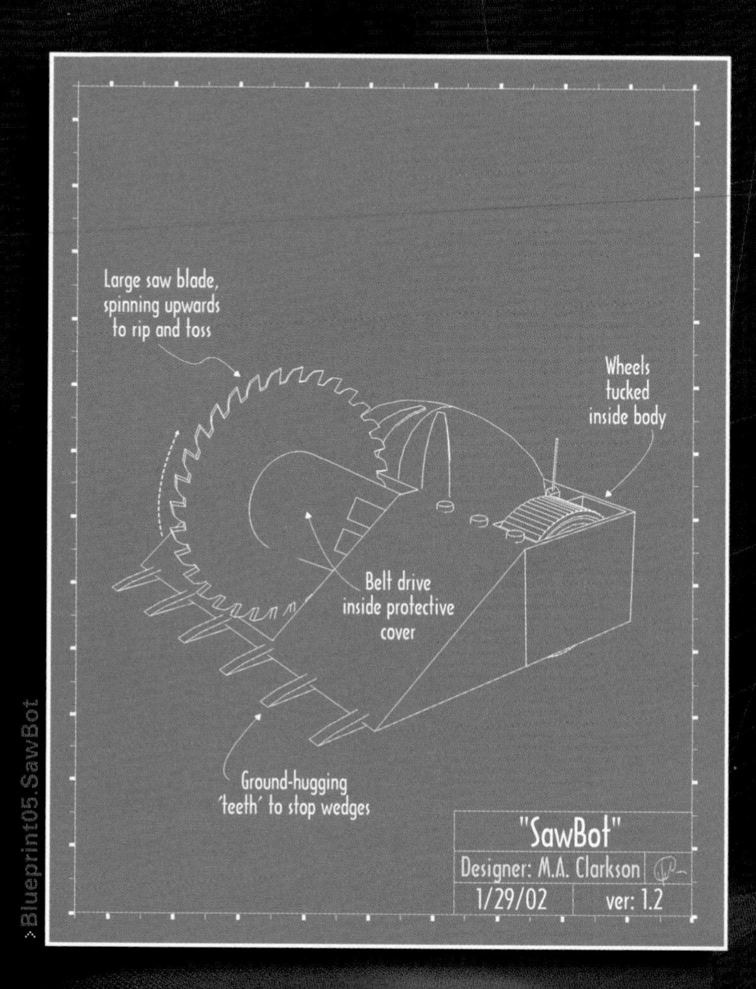

>Blueprint05.SawBot

Turn-ons: Titanium and steel-clad opponents, which make the most impressive sparks. While rarely fatal to the opponent, a powerful saw and the cosmetic damage it leaves can impress the audience and judges enough to give you the win in a close match.

Turn-offs: Well-armored opponents that just won't hold still.

Hobgoblin LTD

While rarely fatal to the opponent, a powerful saw and the cosmetic damage it leaves can impress the audience and judges enough to give you the win in a close match.

> THE LIFTBOT AND LAUNCHBOT

Lifters and Launchers both employ an articulated lifting device (usually an arm). Its purpose is to slip beneath the opponent and lift it into the air, either a little way so that it loses traction and control, or far enough to flip it onto its back.

From a functional standpoint, the primary difference between a LiftBot and a LaunchBot is the speed of the lift. A lifting arm may take two or three seconds to lift its opponent into the air. A launching arm, on the other hand, operates in a fraction of a second. Think of it as the difference between a firm push and a punch in the nose.

The lifting arm is driven by hydraulics, pneumatics, a geared electric motor, or an electric linear actuator. It has enough power and leverage to tilt or lift up the other robot.

Pneumatics are generally much faster, and speed is *de rigeur* for a Launcher. Pneumatics are also easier and cheaper to work with than hydraulics.

The arm can also serve as a self-righting device for the robot, should it find itself on its back during a bout. If the arm extends far enough, or fires quickly enough, it can flip the bot back onto its wheels to continue the fight.

Success with a LiftBot or LaunchBot depends on being able to get the arm seated under an opponent firmly enough to lift. The business end of the arm is often wedge-shaped, or blended into a wedge-shaped front, and in many cases has grip-enhancing hooks or teeth. The arm's pivot point is generally at the back of the robot, for a maximum lift height. The higher the arm raises, the greater the chance that the opponent will be flipped onto its back. A well-designed Lifter can push and drag its opponent freely around the arena. A well-designed Launcher can flip opponents onto their backs or even send them spinning high into the air. Hopefully, something will break when they land.

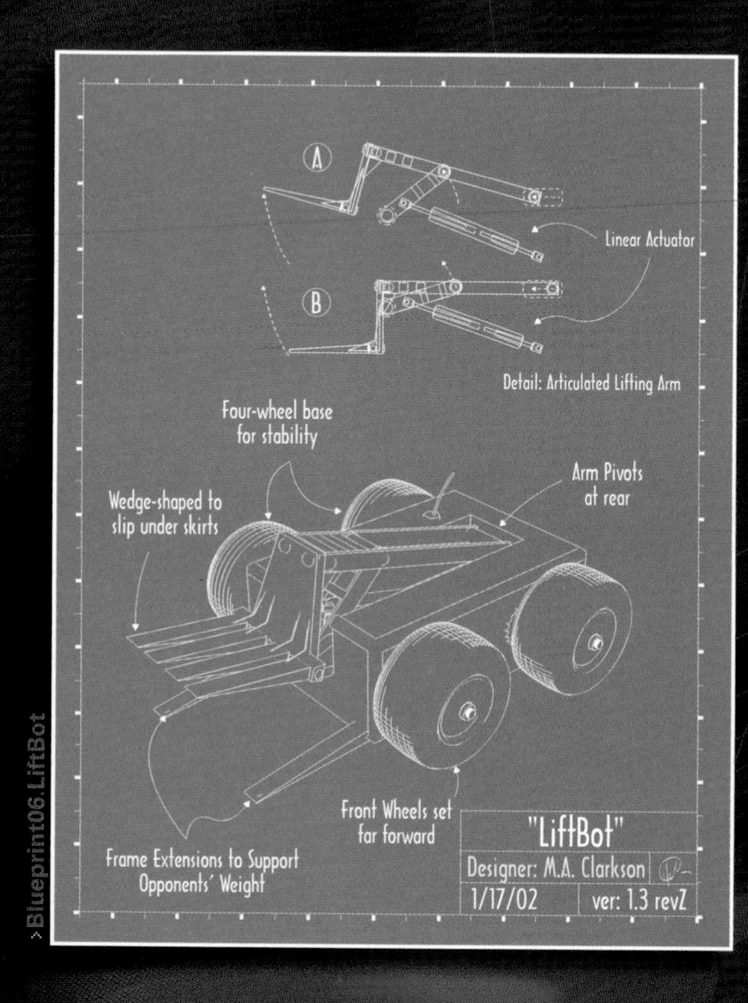

> Blueprint06.LiftBot

Leverage is extremely important to LiftBots and LaunchBots. It does no good to have enough power to pick up an opponent if your bot tips over. The best lifting designs have drive wheels as far forward as possible, flanking the lifting arm, to take advantage of the extra traction arising from the opponent's weight.

> Turn-ons: Opponents such as Wedges and RamBots, which rely on traction to fight, and robots with overhanging enclosed shells will be particularly easy targets for a LiftBot because it can immobilize them by simply tilting one side up high enough to lift their wheels off the ground.

> Turn-offs: SpinBots, ThwackBots, and robots with skirts that are hard to get under. Many SpinBots have enough kinetic energy in their shell to knock the Lifter aside on contact. ThwackBots are also very tough opponents, as their wild spinning and invertible, open-wheeled design make it very difficult for a Lifter to get into position for a lift.

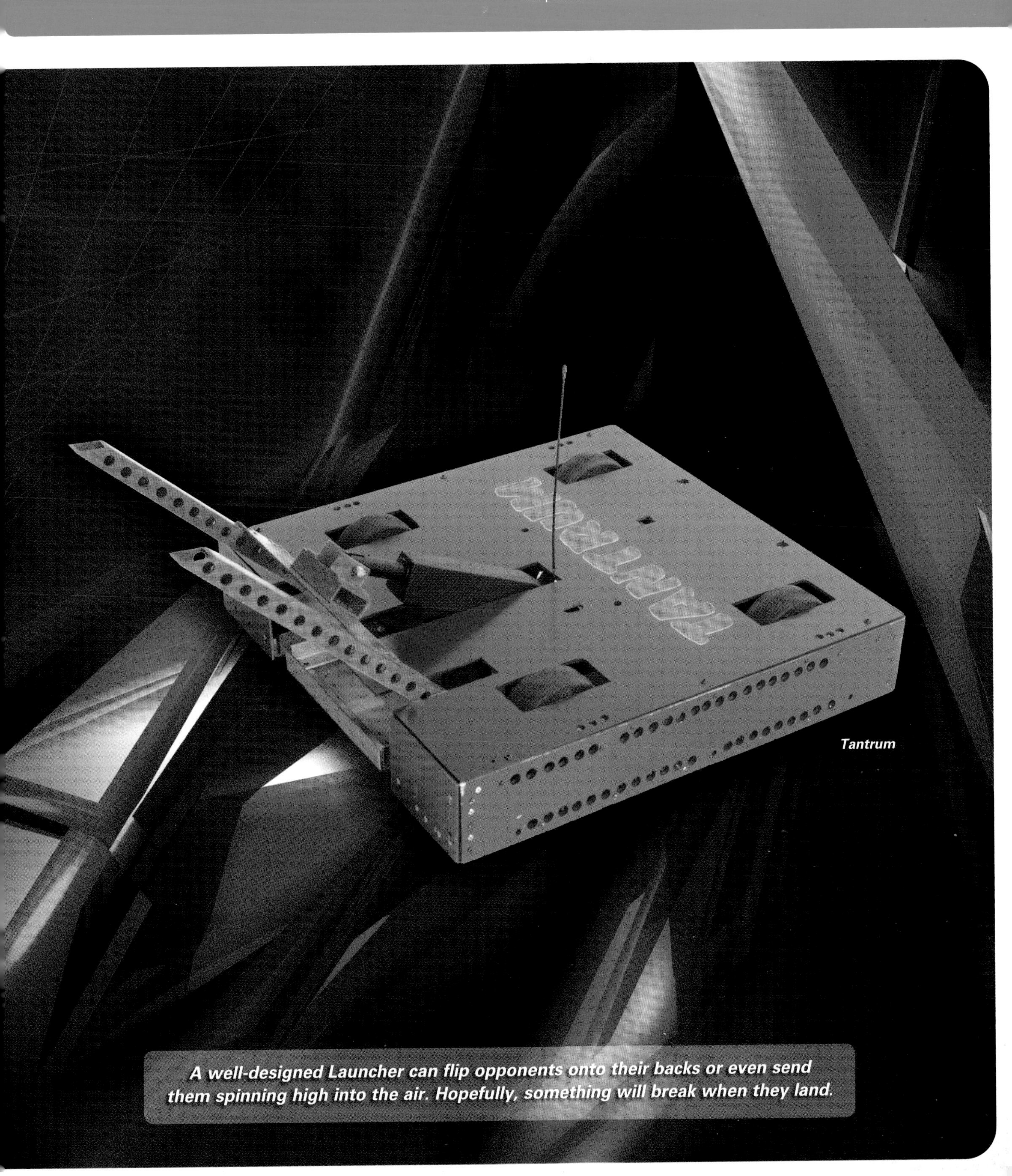

Tantrum

A well-designed Launcher can flip opponents onto their backs or even send them spinning high into the air. Hopefully, something will break when they land.

> THE CLAMPBOT

The ClampBot's strategy is to reach out and grab someone, taking control of its opponent. ClampBots employ a wide variety of jaws and claws, from clamp-shaped bodies to giant maws. Some are meant to grip and hold, others to crush and destroy.

A ClampBot's jaws must open wide enough to grasp its largest opponents, and they should be designed to close before the opponent can escape them. Pneumatics are a popular choice for the closing mechanism, as they can provide both high closing speeds and a strong clamping force. Electric linear actuators or hydraulics are also used. These provide superior closing force at the cost of a slower closing speed.

A ClampBot's drive train should be strong enough to carry at least twice its weight, and the ClampBot's frame must ride smoothly on the ground, even when it's holding onto another bot. The center of gravity should be as low and as far away from the jaws as possible.

One type of ClampBot is really a sub-species of Lifter. While Lifters shove an arm under an opponent and lift its wheels off the ground, this ClampBot adds another component—sort of an opposing thumb—which allows the ClampBot to grab onto its opponents and lift them completely into the air. As very few robots can do anything when lifted off the ground, this places the match completely in the control of the ClampBot.

Some ClampBots, not satisfied with merely grabbing their opponents, or even lifting them into the air, are "Crusher" ClampBots that employ fantastically powerful jaws to actually bite into the other bot, piercing its armor and chewing on its insides. A Crusher's jaws are usually studded with teeth, blades, or spikes to focus the force of the bite onto as small an area as possible, maximizing penetration.

In fact, a crushing ClampBot is one of the few designs that can inflict significant internal damage on an opponent. A Crusher can actually chew holes through an opponent's armor, and into its vitals—radio gear, batteries, and electronics—decisively ending the match.

On the other hand, a crushing ClampBot is probably the most mechanically challenging BattleBot to build. The jaw mechanism must be strong enough to bite through BattleBot armor without collapsing or breaking its teeth. A Crusher needs to achieve both tremendous forces to pierce opponents' armor and quick closing times to sink teeth into an opponent before it can back away. The combination of high force and high speed generally calls for complex hydraulics and a very powerful motor to drive them.

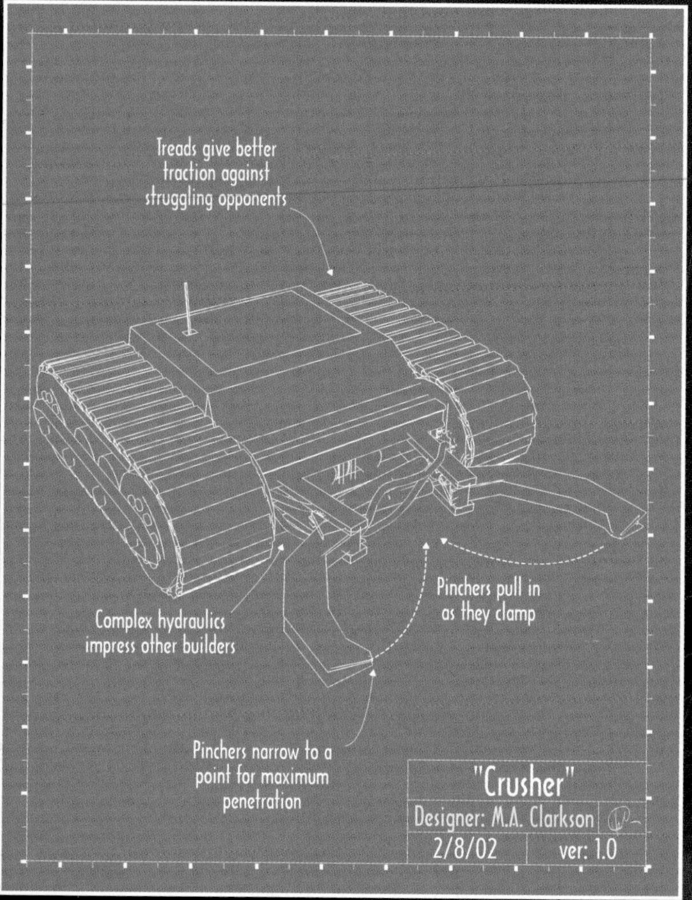

A close relative of the ClampBot is the SmotherBot. Where ClampBots control opponents by clamping onto them with jaws, SmotherBots control opponents by enclosing them in a cage or shell, lowered from above. The SmotherBot can then inflict damage at will with its own weapons, or drag opponents into the arena hazards. The challenge with a SmotherBot is to create a cage that's light enough to lift easily, yet large enough and strong enough to contain an angry BattleBot.

Treads give better traction against struggling opponents

Complex hydraulics impress other builders

Pinchers pull in as they clamp

Pinchers narrow to a point for maximum penetration

"Crusher"
Designer: M.A. Clarkson
2/8/02 ver: 1.0

>Blueprint07.CrusherClampBot

Turn-ons: RamBots, Wedges, and ThwackBots (if it can catch them) — those robots that are completely dependent on their drive power for weapons. Once grabbed and lifted, they are completely helpless. Against a ThwackBot, the challenge for a ClampBot will be in catching its opponent in the first place because many ThwackBots are fast enough to make catching and grabbing them very difficult. Like the RamBot and Wedge, the ThwackBot, once grabbed and lifted, is completely helpless.

Turn-offs: SpinBots, PoundBots, and ChopBots. The SpinBot's weapon must be stopped before the ClampBot can grab it, but the only method the ClampBot has of stopping the shell is by repeatedly ramming it, taking punishing blows before the SpinBot is slow enough to be grabbed. With more working parts and typically lighter frames, ClampBots are more likely than most robot types to be damaged by this kind of punishment. A RotaryBot or BarrelBot is an easier target, if the ClampBot can outmaneuver it and grasp it without taking a hit. PoundBots and ChopBots pose a special problem to ClampBots. A firm grasp can give them the leverage to hit the ClampBot quite hard in the same spot repeatedly.

Fang

A crushing ClampBot also has the advantage that, once it has a victim securely in its jaws, it can take its time to chew and rend. The other bot isn't going anywhere.

THE THWACKBOT

Similar in some ways to the SpinBots and Rotaries, the ThwackBot uses rotational energy to attack its opponents. But where the weapons on those SpinBots and RotaryBots have their own power source, the ThwackBot relies on its main drive wheels to generate spin.

ThwackBots sport a unique, instantly recognizable look. Invariably of two-wheel design, ThwackBots resemble an old-fashioned lawn mower, with a pick, hammer, or axe where the handle would be. When the driver starts the two wheels turning in opposite directions, the ThwackBot spins in place, whipping its weapon around at high speed on its long tail, creating a "circle of death" of significant size and power. Some ThwackBots use two arms, one on each side, for balance and to deliver twice as many blows.

The length of the ThwackBot's tail gives it much better leverage than a SpinBot; the farther from the bot's center of spin, the faster the weapon moves at a given rate of spin. ThwackBots' weapons typically weigh much more than SpinBots', as well, but ThwackBots can't rotate nearly as fast as SpinBots.

ThwackBots can usually run perfectly well inverted, making them a difficult opponent for Wedges or Lifters.

ThwackBots are fairly easy to design and build, but they do pose a pretty problem in balance. The more weight placed in the weapon (at the point of impact), the more damage the weapon will do. At the same time, any weight resting on the tail or on any idler wheels deprives the bot of the traction it needs to spin up quickly.

Ideally, a ThwackBot should be able to reach top rotation speed in less than a single revolution, yet still have a top speed fast enough to do damage on impact. A ThwackBot that takes too long to spin up will find itself helpless once an opponent has come to close range and survived one hit. Of course, more power makes for shorter spin-up time, higher top speed, and more damage done.

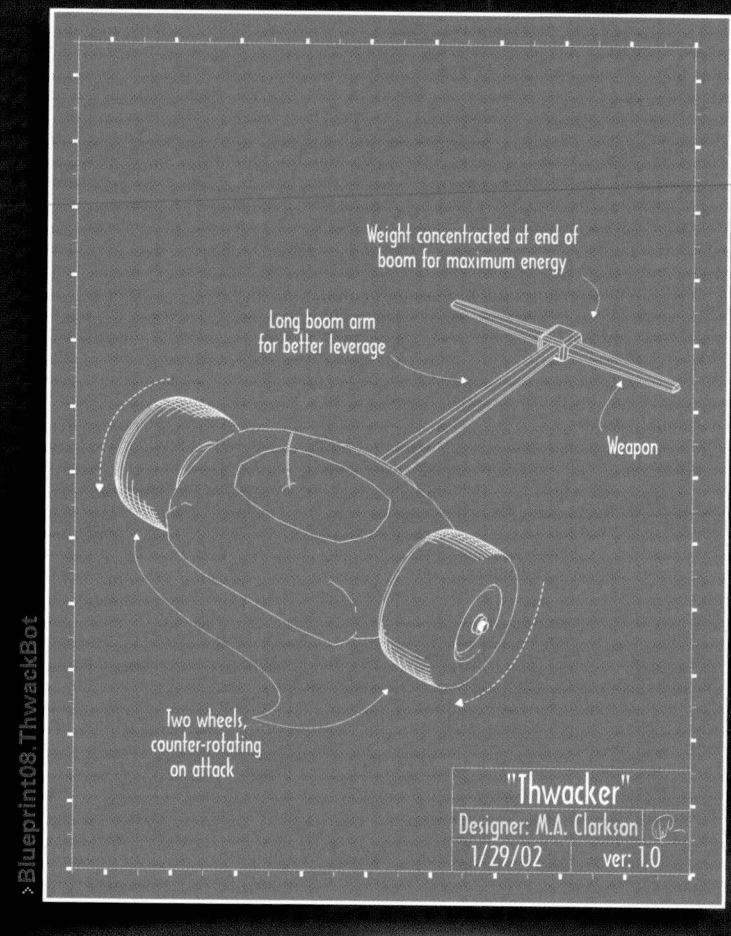

>Blueprint08.ThwackBot

The ThwackBot suffers from a unique disadvantage: it can't steer (or, in fact, even move) when its weapon is spinning. With its two wheels turning full speed in opposite directions, the ThwackBot is compelled to sit and spin in one place and hope its opponent comes near enough to take some damage. There's been a lot of discussion and experimentation among BattleBot builders about the design of a so-called "Melty Brain" system. The Melty Brain is a computer controller that keeps precise track of the ThwackBot's orientation and varies each wheel's velocity slightly, over fractions of a second, to allow the ThwackBot to turn and move even while spinning its weapon.

> Turn-ons: Lifters, Launchers, and Grabbers. A strong spinning attack can keep these bots from sliding their arms underneath, and the open-wheeled design and powerful drive of most ThwackBots makes them difficult to keep a grip on.

> Turn-offs: Wedges, BarrelBots, and RotaryBots. Wedges are difficult opponents for a ThwackBot. Victory often comes down to speed and maneuverability. Barrels and Vertical RotaryBots can also be very dangerous customers, as they can catch the ThwackBot's long weapon boom and toss it violently upward.

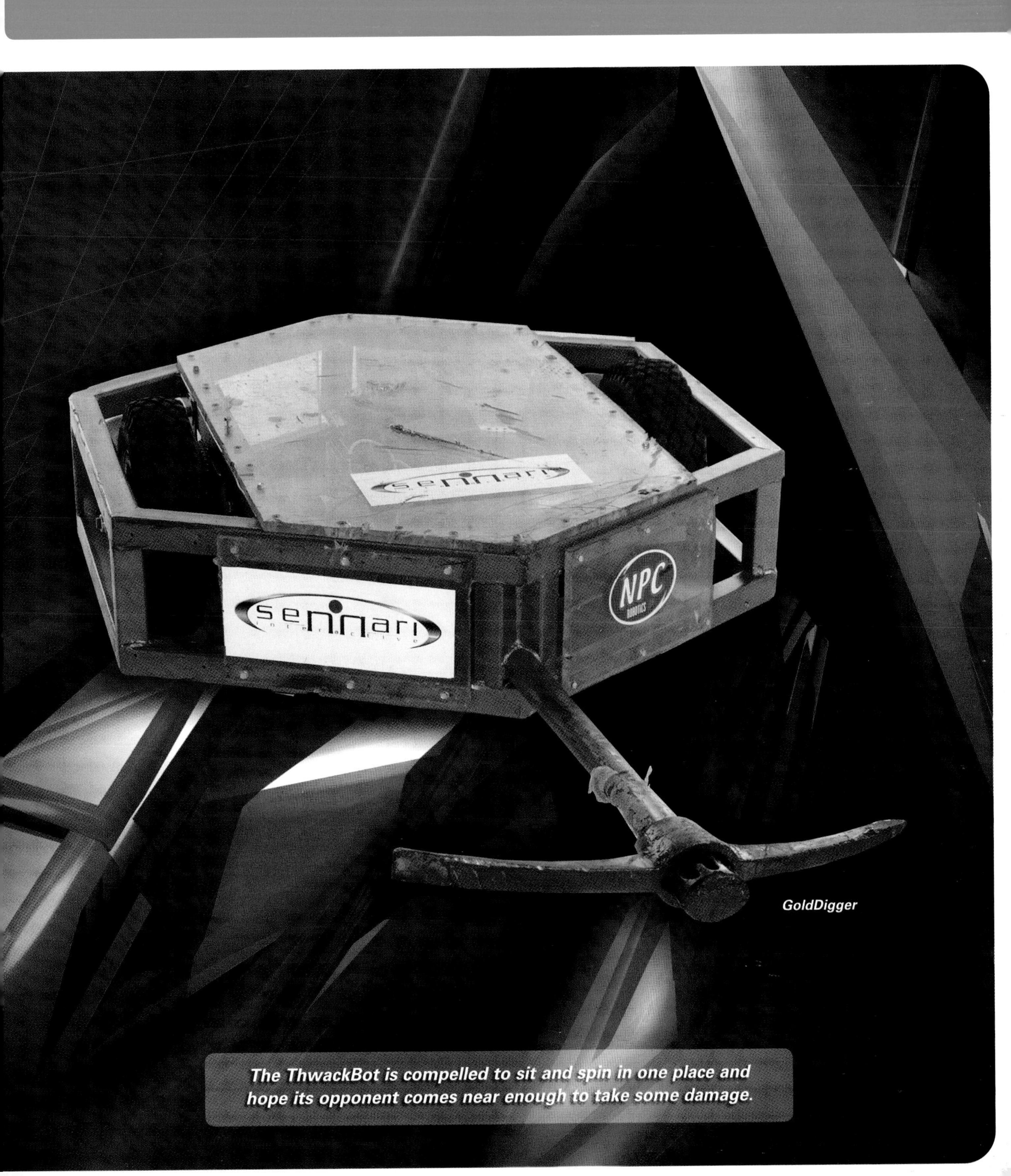

GoldDigger

The ThwackBot is compelled to sit and spin in one place and hope its opponent comes near enough to take some damage.

THE POUNDBOT AND CHOPBOT

PoundBots and ChopBots attack their opponents by hitting them on the head with hammers and picks (PoundBots) or blades and axes (ChopBots).

Unlike ThwackBots and SpinBots, which can take a few seconds to build up speed and kinetic energy, PoundBots get one swing of, at most, about 180 degrees, to dump as much energy as possible into their weapon before it hits their opponent. Consequently, PoundBots and ChopBots have a much harder time inflicting real damage on other bots.

This disadvantage is offset by this design's ability to control the timing and placing of its hits, to strike repeatedly in a short period of time, and to use its weapon even if pinned or lifted completely into the air. Most hammer weapons can even serve as a self-righting mechanisms if the bot is flipped.

Most PoundBot and ChopBot weapons are pneumatically or electrically driven. Hydraulic power can provide tremendous force, which can accelerate a weapon very quickly or drive a weapon, slowly and inexorably, through even the toughest armor. But hydraulics are complicated and expensive to work with. Some builders have experimented with using a large spring to power the weapon, and a high-torque motor or linear actuator to crank the weapon back and latch it after firing. This can give a very powerful hammer action but a fairly slow reload.

One type of PoundBot uses the bot's drive motors and inertia to power the weapon. As in ThwackBots, these guys mount a weapon at the end of a long arm, attached to a compact, two-wheel-drive chassis. They attack by charging at their opponents at full speed and then, at the last instant, slamming into reverse. This pitches the weapon (and, often, the entire bot) forward and over, 180 degrees, like a motorcycle rider hitting a wall. If the timing's right, the weapon slams down right on top of the opponent.

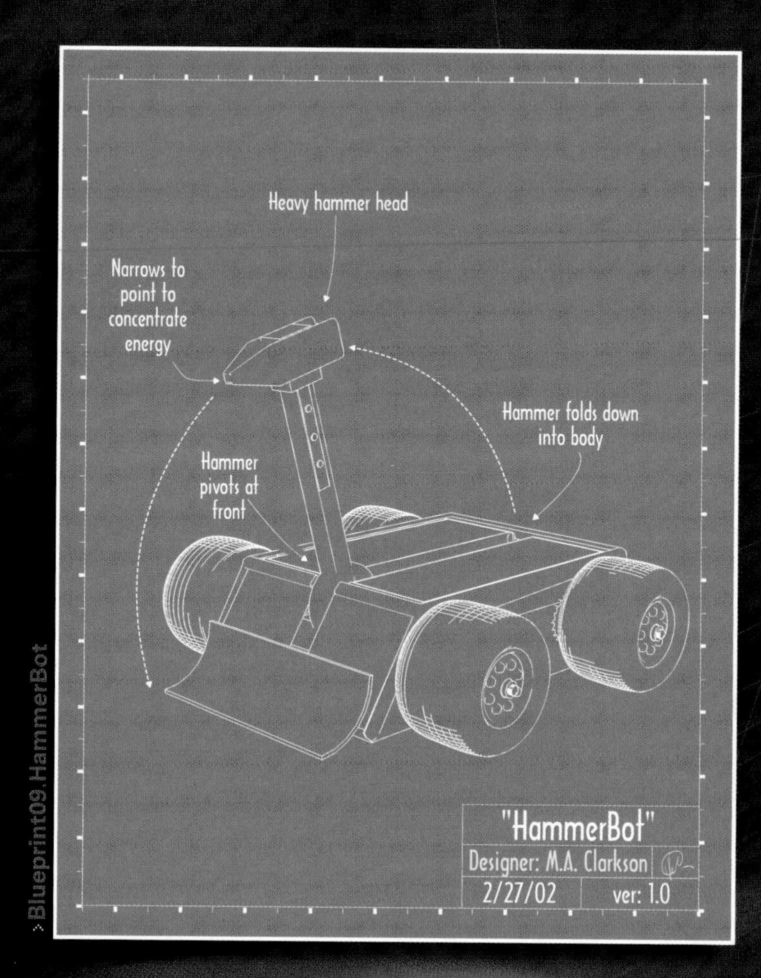

> Blueprint09.HammerBot

Even the strongest Pounders and Choppers have trouble consistently disabling opponents with their weapons. They can win matches, however, by being more aggressive and delivering lots and lots of weapon hits in the course of a match, racking up points.

Turn-ons: Robots that hold still.

Turn-offs: SpinBots, agile robots, and heavy armor. Striking the active SpinBot, anywhere, is likely to result in a bent or lost weapon. PoundBots and ChopBots usually have weapons with very small strike zones; agile robots especially can contrive to stay out of it. Even if they make contact, PoundBots and ChopBots have a hard time putting sufficient energy into their hits to do any damage to a robot with heavy armor.

Overkill

Hydraulic power can provide tremendous force, which can accelerate a weapon very quickly or drive a weapon, slowly and inexorably, through even the toughest armor.

THE BARRELBOT

BarrelBots typically have four wheels and a roughly square overall shape, with a wide spinning barrel or drum, usually bristling with spikes or teeth, mounted on the front. BarrelBots use their weapon to grind opponents into submission.

The BarrelBot's nice, thick barrel drinks up more damage than a relatively thin blade or disk. It's usually large enough for the rest of the BattleBot to hide behind, letting the BarrelBot take all its opponent's hits where it's strongest.

BarrelBots can hit an opponent repeatedly in a short period of time. With its lower center of gravity and less gyroscopic effect to fight, it can be faster and much more nimble than a Vertical SpinBot. The wide barrel doesn't need to be aimed very carefully to be effective.

As the impacts from the Barrel tend to lift opponents into the air, the BarrelBot functions well as a RamBot, repeatedly kicking its opponent across the arena with a combination of weapon hits and drive power.

Like a Vertical RotaryBot, the BarrelBot is subjected to a major downward impact every time it strikes an opponent. Most BarrelBots place the front wheels well forward, and even add support arms or wheels under the Barrel weapon to support it on impact and keep the front end from being driven into the arena floor.

Fast, maneuverable, and with fast weapon spin-up times, Barrels lend themselves to an aggressive driving style. BarrelBots can take control of the match early and keep the opponent on the defensive.

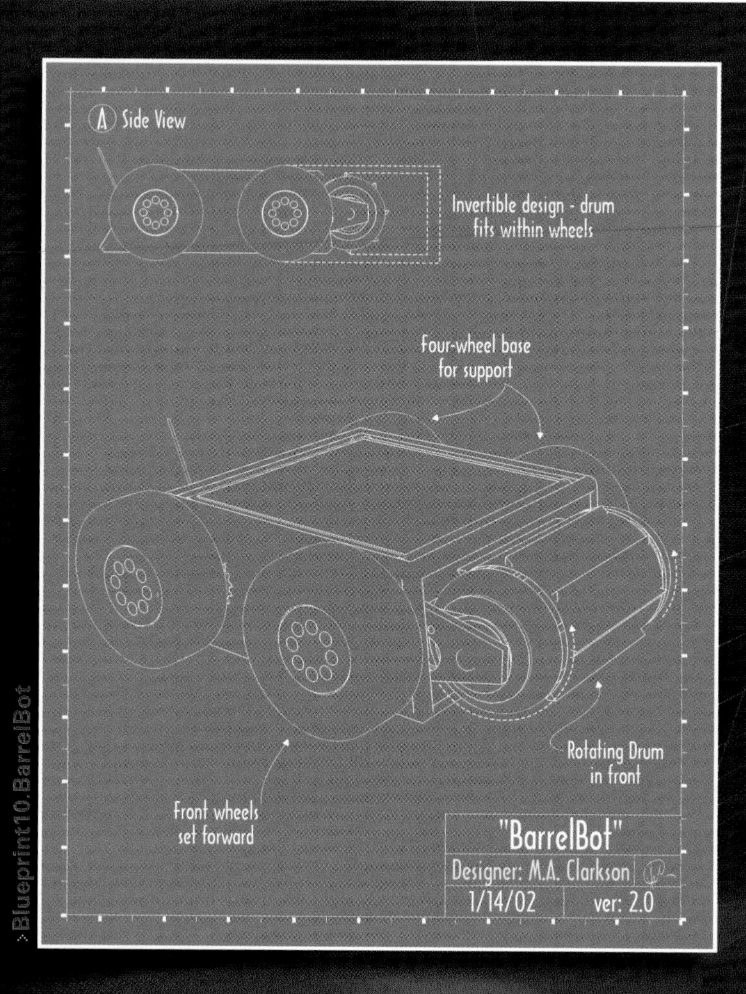

>Blueprint10.BarrelBot

Turn-ons: Slow opponents. A good BarrelBot can usually beat BattleBots that need time to spin up, set up controlling moves, or inflict significant damage.

Turn-offs: Wedges and SpinBots. The Wedge is the bane of the BarrelBot. A Wedge's sloped front doesn't offer a good surface for the Barrel's weapon to catch on. A fight between a BarrelBot and a SpinBot will usually hinge on whether the Barrel's weapon drive and support structure can hold together long enough for the SpinBot to be disabled.

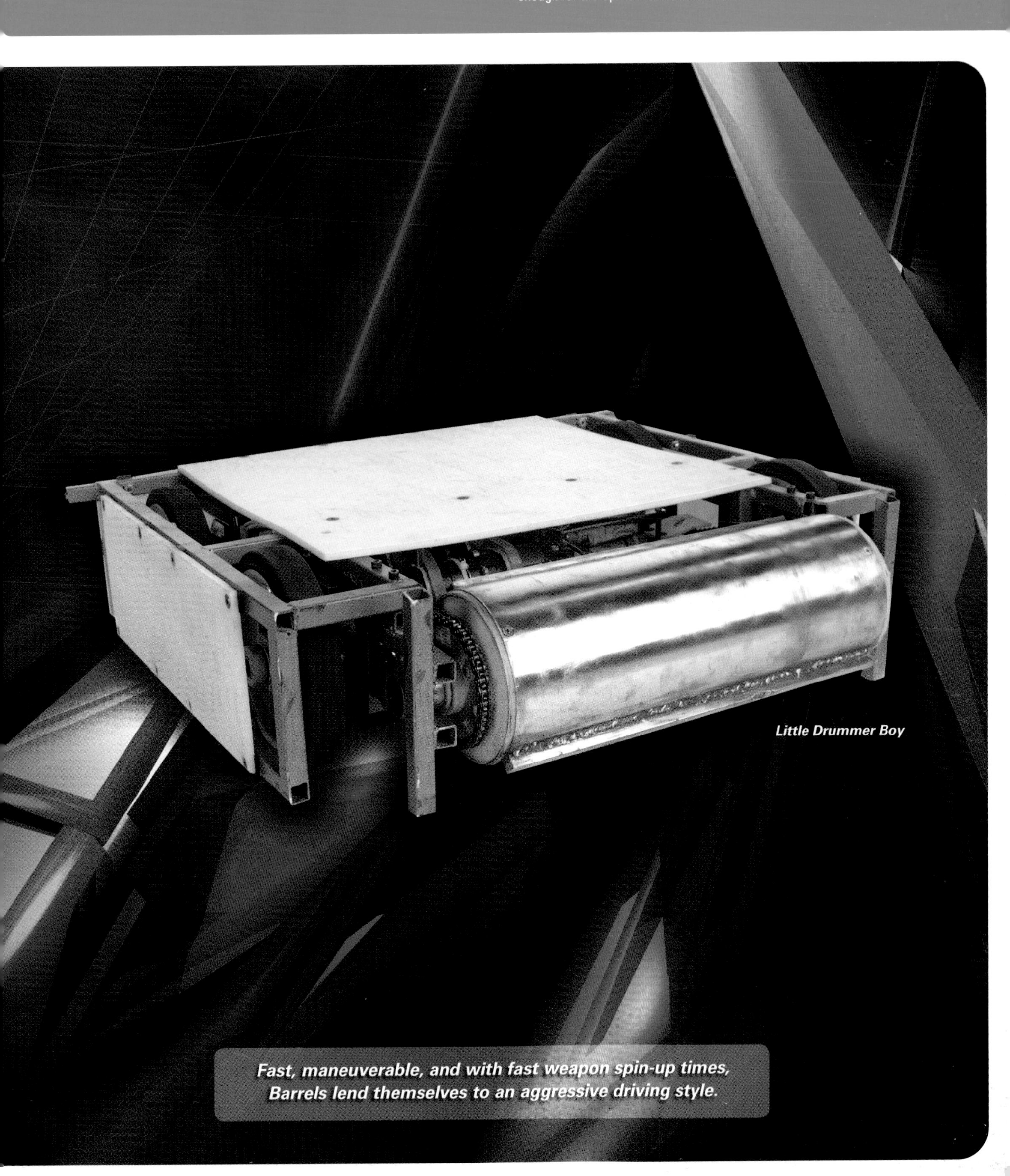

Little Drummer Boy

Fast, maneuverable, and with fast weapon spin-up times, Barrels lend themselves to an aggressive driving style.

> THE PUNCHBOT

Think of the PunchBot as a mobile gun platform. The PunchBot fires a spike or ram at its opponent, hopefully inflicting damage or at least knocking it back. The PunchBot's weapon is almost always pneumatically powered, allowing the weapon to be accelerated very quickly, like a shell leaving a gun. The weapon doesn't actually detach from the robot—BattleBots rules prohibit the firing of untethered projectiles—but the effect on opponents within range is nearly the same. And a PunchBot's weapon can sometimes extend out as far as several feet, allowing it to "reach out and touch" opponents from a comparatively safe distance. (As a general rule, the farther out the weapon extends, the less force it'll hit with.)

PunchBots are generally sturdy bots with at least four wheels, to provide a solid platform from which the weapon can be fired.

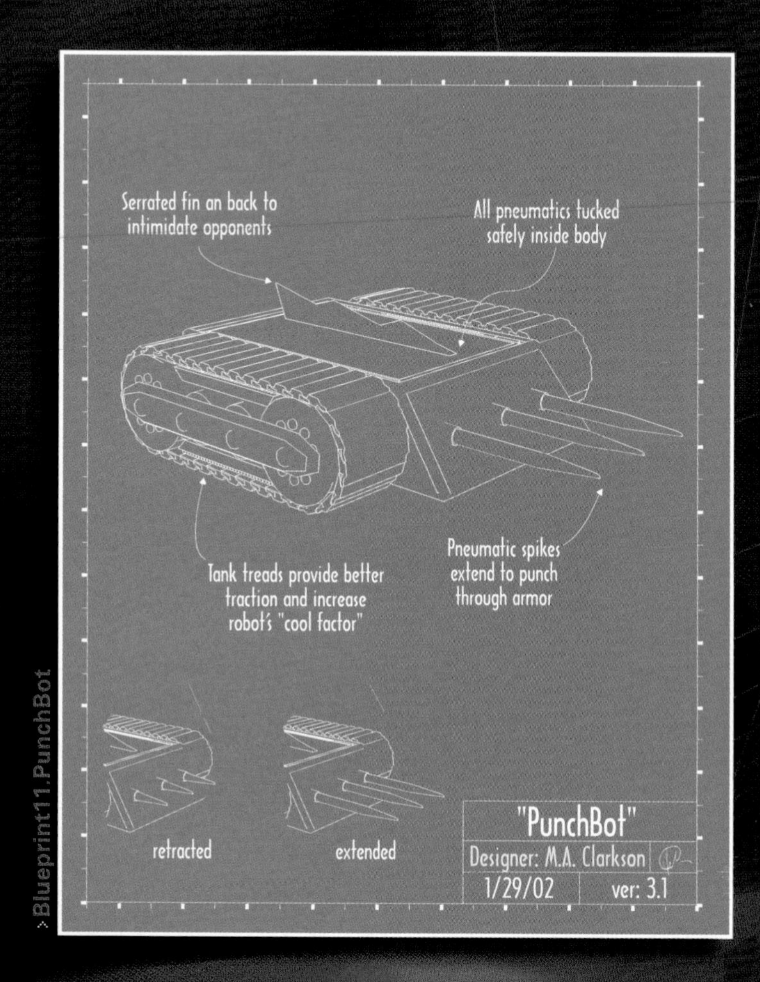

> Blueprint11.PunchBot

Turn-ons: Opponents with lots of relatively flat surfaces to punch at, and the taller the better.

Turn-offs: Wedges, ThwackBots, SpinBots, and really short robots. Extremely low robots might be able to duck underneath the weapon, like a soldier crawling on his belly while shots fly overhead. PunchBots have a difficult time reaching a ThwackBot, and a punch is nearly useless against a SpinBot. The punching weapon is as likely to be torn off as to do any real damage. The weapon will tend to glance off a Wedge, perhaps even lifting the PunchBot off the ground as the weapon rides up the slope. If, on the other hand, it can catch a Wedge on a flat side or an exposed wheel, it can strike an effective blow.

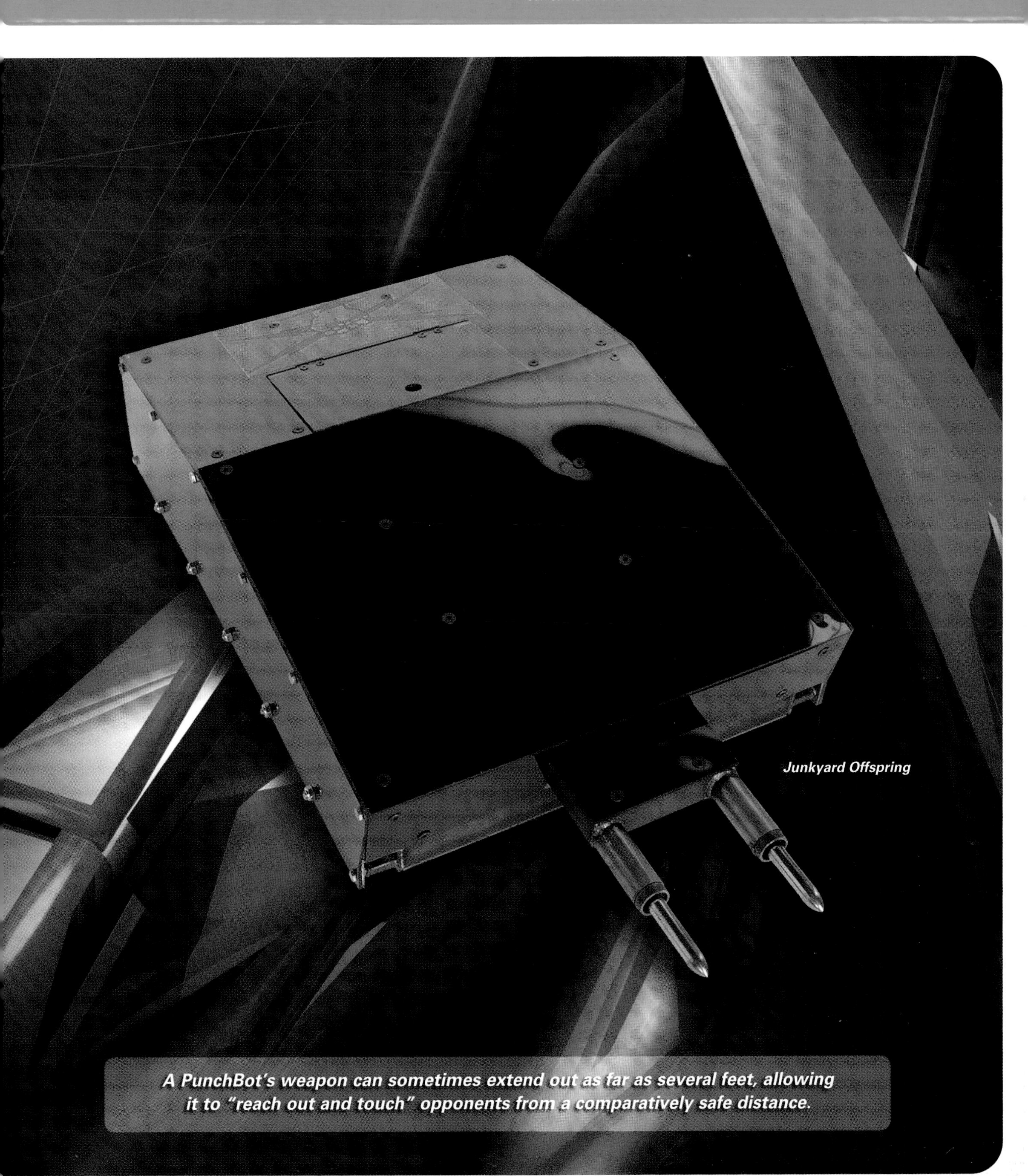

Junkyard Offspring

A PunchBot's weapon can sometimes extend out as far as several feet, allowing it to "reach out and touch" opponents from a comparatively safe distance.

THE MULTIBOT

If you don't find the thought of designing, building, and driving one BattleBot to be sufficiently intimidating, the BattleBots rules provide for the construction of "MultiBots": BattleBots that can split into two or more independently operating bots. Very few MultiBots have been fielded in BattleBots thus far, but they offer some intriguing possibilities.

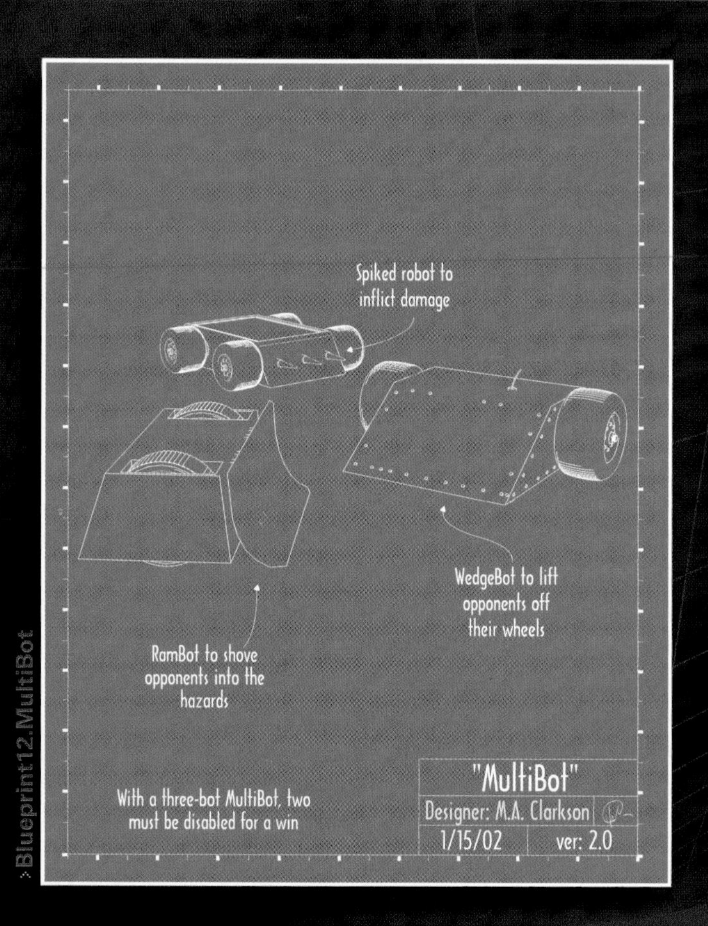

>Blueprint12.MultiBot

MultiBots can attack from multiple directions at once, ganging up on opponents. There are those who find MultiBots to be, well, not entirely sporting. Once a MultiBot breaks apart, it's two against one (or even three—or more—against one) and that ruffles some fans' feathers. But the advantage in numbers is offset by a comparable disadvantage in weight. A single 170-pound Heavyweight might split into, say, two 85-pound Middleweights. There's two of them, but each one is facing an opponent twice its own weight. And the rules state that when 50 percent of the MultiBot, by weight, is disabled, the MultiBot loses.

MultiBots may employ any weapon or combination of weapons, and any strategy or combination of strategies. As weight is such a big factor in all things BattleBot (heavier = more powerful motors; heavier = thicker armor; heavier = more massive weapons), MultiBots are at an inherent disadvantage going toe to toe with almost any kind of opponent. Any weapons they encounter are likely to be larger and more powerful than their own.

On the other hand, they're likely to be more nimble than their larger adversaries. And individual MultiBots can afford to specialize—one bot holding the opponent immobile or at least keeping it occupied while the other flits around to nip at its heels.

Turn-ons: Big, slow robots with big, slow weapons. Being smaller and hopefully quicker, MultiBots can attack slower opponents from different directions and dodge out of the way of their weapons. Even if they can't disable opponents, they can hope to get in more hits, act more aggressively, and win on points.

Turn-offs: Solid, well-armored robots, low to the ground with lots of traction. Since each MultiBot segment weighs only half or a third (or less) of what their opponents weigh, a well-armored opponent with plenty of traction can prove impervious to their assault, drinking up hits without taking damage, and holding its ground without letting itself be shoved around.

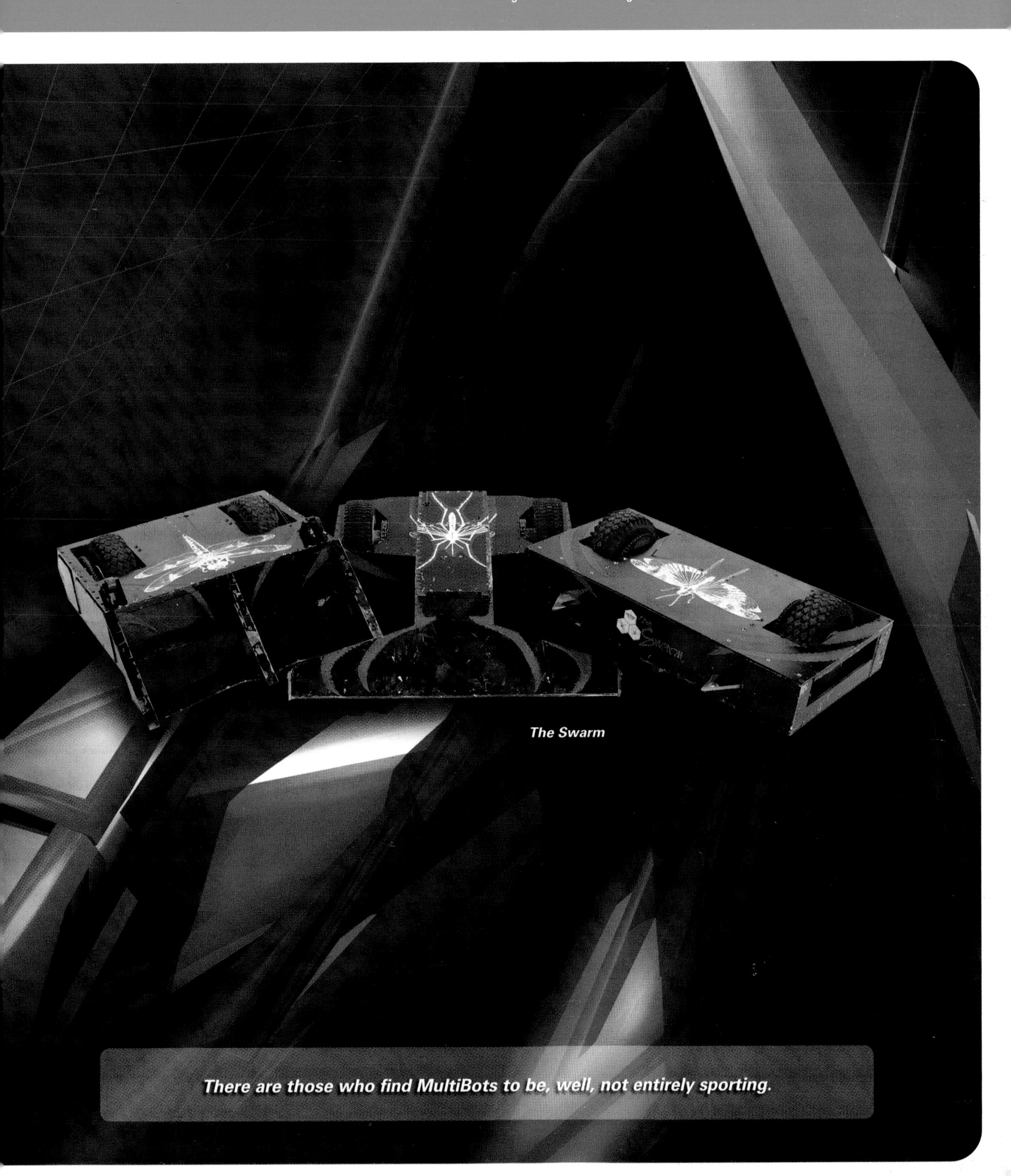

The Swarm

There are those who find MultiBots to be, well, not entirely sporting.

THE "CHIMERA"

Chi' • me • ra *n. 1. A fire-breathing she-monster, part lion, part goat, and part serpent. 2. A monster made up of grotesquely disparate parts.*

For all of my easy talk of dividing BattleBots up into different types, the real world is more complicated than that. Builders are not limited to bots that fit neatly into one category. Many, if not most, BattleBots are actually a combinations of types. Wedges with saws. ClampBots with hammers. Wild combinations of every weapon type the builder can figure out how to mount on the same chassis. Look closely and you'll see that the majority of BattleBots, no matter what the type, incorporate a wedge or two into their designs somewhere.

Builders enjoy taking inspiration from an existing design and then adding a little twist of their own. That twist often takes the form of a new or additional weapon: "What if I built a ThwackBot with a wedge-shaped hammer?" or "What about a Wedge with a built-in lifting arm?" or "Spikes! I'll add spikes!"

Some robots even come with interchangeable weapons, which can be swapped in and out as the occasion warrants. They might wield a sword for one match and a hammer in the next; the team is free to choose the weapon best suited to each opponent. A relatively fragile weapon might be replaced with a simple, sturdy plow blade for a bout against a destructive SpinBot. Bots with multiple interchangeable weapons must be weighed in each configuration they'll be competing in, to make sure the robot remains within its weight class limits no matter which weapons it's wearing.

The individual components of MultiBots can each employ different weapons and strategies, as well, bringing to bear whatever combination of weapons suits their builders. Moreover, the individual MultiBot segments can themselves have interchangeable weapons.

As long as none of the weapons are on the "forbidden" list (no machine guns, no burning oil slicks, no using lasers to blind the other driver), if you can fit it on your bot, you're free to bring it into the arena and try to mete out punishment with it.

Vermicious Kenid

>Turn-ons and Turn-offs: Chimeras defy categorization. You can say Wedges are good against RamBots, and Spinners are death on ClampBots, but a Chimera might be anything. A Spinner and a Wedge. A BarrelBot and a Rotary. A ClampBot with a Hammer. A ThackBot with interchangeable weapons. The Chimera combines the strengths and weaknesses of the different types—perhaps doubling up on strengths, or canceling out a weakness. By definition, you can't say what it's good or bad against, because you can't say what it is.

> From Platonic Ideal to Motorized Reality

So those are the types of BattleBots, a basic vocabulary for use when discussing BattleBots with friends and family ("Wedges are for wimps!" "No, they're not, Grandma!"). But to say that all BattleBots fall into one of a handful of categories is to belie the incredible variety the robots display. It's like saying that all country songs are written with the same twelve notes: true but misleading. I've said it before and I'll say it again: no two BattleBots are alike. Each is the mechanical manifestation of its builders' unique collective vision. Each, in its own way, is singularly beautiful. *(sniff)*

Diesector

>Props: Some of the material in this chapter is courtesy of Andrew Lindsey, and appears in a different, more comprehensive, form in *Build Your Own Combat Robot,* by Pete Miles and Tom Carroll (McGraw-Hill/Osborne, 2002). *Thanks, Andrew!*

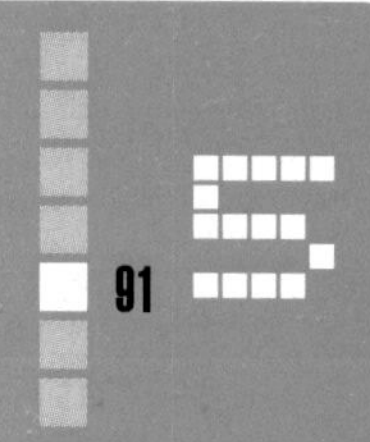

THE ROBOTS

Sure, one can talk about the different types of robots: Spinners and Wedges and Rammers and Hammers. But the actual robots themselves are not so easily pigeonholed. Each robot is a unique creation, with a unique personality. No two robots, not even two Lightweight Wedges, are identical in design, construction, or driving style. They display the same wild variety as the builders who spawned them.

Dogs are said to mirror the appearance of the people who own them, but BattleBots reflect that part of their builders' psyche that rarely gets to come out and play. Every builder puts unique twists and turns into his or her robot. It is most definitely their own creation and no one else's.

Come with me, if you dare, and meet a few of the robots that have made names for themselves, and their builders, in the fearsome sport of BattleBots.

Dogs are said to mirror the appearance of the people who own them, but BattleBots reflect that part of their builders' psyche that rarely gets to come out and play.

AFTERTHOUGHT

Afterthought takes it on the chin

"I've always been a tinkerer," says Afterthought's builder, Jim Sellers. "I loved building mechanical and electronic devices, gadgets, and gizmos." Sellers has worked as a production engineer, a design engineer, electromechanical technician, a machinist, a welder, and a furniture maker. Then, one fateful day, a coworker invited Sellers to help build a robot for a robot fighting competition. "I helped, it won, and the rest is history!" says Sellers. "I was hooked."

Sellers' girlfriend is supportive of his addiction, or at least reconciled to it. "I think she has resigned herself to the fact that during the last couple of weeks before the bot gets shipped, helping is the only way she gets to see me!"

Sellers says Afterthought's current design was inspired by destructive spinning robots such as Ziggo and Nightmare. "I realized just how much energy could be stored in a rapidly rotating mass... a lot!"

Cardboard-Aided Design

Sellers' design methodology is largely low-tech: "I'll use CAD or a pencil and paper for machined parts that have to fit closely, that's about it. I've always been a cardboard and wood mockup kind of guy."

Originally built around a leftover frame from another robot, Afterthought's current chassis is made from two pieces of 6061 aluminum channel, tied together with (stronger) 2024 aluminum crosspieces, mounted on a base plate made of (stronger still) 7075 aluminum. There are two pieces of 1/16" titanium plate inside the frame under the battery packs and speed controls to provide a little extra protection from the Kill Saws for the more expensive components.

Afterthought is four-wheel-drive, using a combination of belts and chains. Two EV Warrior electric bicycle motors are powered by 24-volt NiMH BattlePacks. "It's the motor of choice for bot builders on a budget," says Sellers. "You won't get better for under $20! I ran them at 24 volts, twice their rated voltage, for four times the power output."

Afterthought's weapon is an 18-inch diameter, 13-pound disk—part of a cultivator, picked up at a local farm store. Sellers had the disk laser cut to true it up, and punched some holes in it to reduce the weight. It's spun up to 2500 RPM—134 MPH at the tips of the blade's cutters—by dual Astroflight Cobalt 40 motors driving through Sellers' custom-built gearbox.

More Armor, Please

Afterthought has been through three revisions, acquiring new drive motors, new speed controllers, a new chassis, and a bigger weapon. "Actually," says Sellers, "I don't think there is any of the original robot left, except the radio and one of the weapon motors!"

The robot will undergo another revision before the next competition. "I'll probably replace the chains with belts," says Sellers, "and go back to the original disk. It's slightly smaller, but I can spin it faster with better cutters. And definitely better armor, possibly Kevlar panels." The 1/8" polycarbonate armor on the last version of Afterthought was, well, a little disappointing, Sellers admits: "Okay, it was pitiful! But I only had five pounds left for armor."

Bot Name: Afterthought
Weight Class: Lightweight
Team: Team Afterthought (part of Robot Action League)
Jim Sellers
"I realized just how much energy could be stored in a rapidly rotating mass ... a lot!"
— Jim Sellers

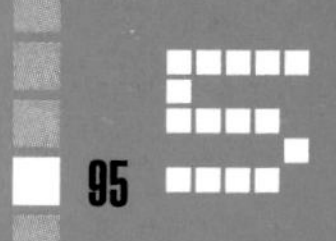

ATOMIC WEDGIE

"When I saw the first episode of BattleBots on Comedy Central, I lost my mind!" says Team Half-Life's Rob Everhart. "I got all excited like a kid seeing a new toy on TV." Within 24 hours he had enlisted the other members of his team—his coworkers at a nuclear waste clean-up company—and begun designing a BattleBot.

Atomic Wedgie gets under Techno Destructo

Why a Wedge?

The team initially designed a SpinBot with two counter-rotating halves, but it soon became clear just how complicated and expensive that was going to be. They needed a new, simpler design. They eventually decided on a Wedge: proven, robust, and relatively easy to build. Within two months, Atomic Wedgie was ready to roll.

The two spinners survived in the form of dual 15-pound tri-foil weapons, mounted at the rear of the bot and spinning at nearly 150 MPH. For Season 4.0 the team switched to a single aft-mounted blade, which weighs about 30 pounds and spins in the neighborhood of 300 MPH.

The Wedgie's body is welded 6061 aluminum plate, covered with 1/8" or more of titanium armor. "We have never had an armor breech," says Everhart. The Wedgie's ramming nose is made of a super-hard nickel alloy called Inconel, found in nuclear reactors as well as BattleBots.

Tread Traction Troubles

Atomic Wedgie features two sets of custom-made tank treads. One set on the bottom drives the bot, and another set on the top pulls the victim toward the Wedgie's spinning weapon. With treads on top and bottom, Atomic Wedgie can drive inverted. But the dual tank tread drive requires a complex drive train combining gears, chains, sprockets, and six jackshafts.

Atomic Wedgie did very well its first time out, placing second in its weight division. Nonetheless, the team has reworked the robot each season. The robot started with National Power Chair motors, but the gearbox needed to speed the drive train up was a weak link. Next came Bosch 750s, but a combination of errors and circumstance burned them up almost immediately. For Season 4.0, the team redesigned the drive train again. Wedgie currently runs on Magmotors, driven by four Hawker gel cell batteries.

Wedgie's treads have changed as well. In its first incarnation, Everhart says, Atomic Wedgie's rubber-coated treads provided too much traction. "We actually broke a shaft from the abuse we put our drive train through, due to having too much traction. So for Season 3.0 we removed the rubber and made the treads wider... and they switched the floor paint in the arena. We slipped so badly that the motors burned up. For Season 4.0, we're using the wide treads with a new rubber top."

East Coast Boys

All told, Team Half-Life has invested more than $30,000 in the Wedgie, and not all of it in parts. "A ton of our money goes toward travel. Being an East Coast team, we have to budget travel, lodging, and shipping of the bot into the mix."

- **Bot Name:** Atomic Wedgie
- **Weight Class:** Super Heavyweight
- **Team:** Team Half-Life (www.getawedgie.com) — Rob Everhart, Charlie Payne, Tom Corrie, Kevin Maze, and Rich Martin; not pictured, Brad Bowen and Joe Esposito

The team initially designed a SpinBot with two counter-rotating halves, but it soon became clear just how complicated and expensive that was going to be.

BACKLASH

(Back • lash) n: a movement back from an impact.

Skid Mark vs. Backlash

I always think of Backlash as a sort of "little brother" to Jim Smentowski's other signature BattleBot, Nightmare. Although Backlash is much smaller and shinier, the family resemblance is unmistakable. Something about the eyes, I think.

Aluminum, Titanium, and Steel

Like Nightmare, Backlash's salient feature is its large, vertically spinning disk, with two heavy steel "teeth" for delivering vicious uppercuts to its opponents. Backlash's 22-inch diameter disk is spun up to 130 MPH by a powerful electric motor rated at about 2 HP. It can flip bots in the air and tear big chunks off of them. But there's a down-side, says Smentowski. Backlash gets it as good as he gives, absorbing an equal amount of shock from the impact. Backlash has had to be built super-tough to withstand repeated impacts without destroying itself.

Backlash was built from the base plate up, says Smentowski. "I came up with the placement of all the components, laid them on a sheet of aluminum, and built it up from there. I then covered the whole thing in armor. On the current version of Backlash, I'm using 7075 aluminum armor. It's very strong, and, at this thickness, has a cool ability to pop back into shape when impacted."

Backlash's disk is usually machined from hard aluminum. However, at the Season 4.0 event, Smentowski tried out a more expensive titanium disk. "I was having trouble with Backlash's disk," he says. "It was bending and warping on impact. I wanted to see if titanium would prevent that. It did." But when the aluminum disk flexes, it absorbs some of the shock of impact; the stiffer titanium disk transfers more of that shock to the rest of the system. "I'm still evaluating whether I want to keep the titanium or not," says Smentowski. "But it was fun to try it out."

NiCads, NiMHs, and Secret Motors

Backlash's drive wheels are chain-driven off of electric EV Warrior motors—originally built for electric bicycles—running at 14.4 volts. "I used these motors because they are powerful, relatively lightweight, inexpensive, and easy to find," says Smentowski. The weapon's motor? "That's a secret," he says.

Smentowski uses two different types of batteries in Backlash: NiMHs to power the drive, and NiCads to power the weapon. "The NiMHs last longer," says Smentowski, "which is what I want in his drive, but the NiCads can push more amperage, which I need for that weapon motor. It draws a ton of amps when it's starting up or moving through an impact."

Constant Revision

As effective as Backlash is, it's helpless if an opponent or arena hazard tips it over. "Obviously," says Smentowski, "I need to do something about Backlash's inability to operate upside down or to recover from being flipped. I'm re-engineering the whole bot for that purpose."

All told, he says, Backlash has already been through six revisions in three years, some of them minor upgrades and some of them complete rebuilds. "I think at any time I could put together three Backlashes from what I have in the garage."

Bot Name: Backlash
Weight Class: Heavyweight
Team: Team Nightmare (www.robotcombat.com)
Jim Smentowski

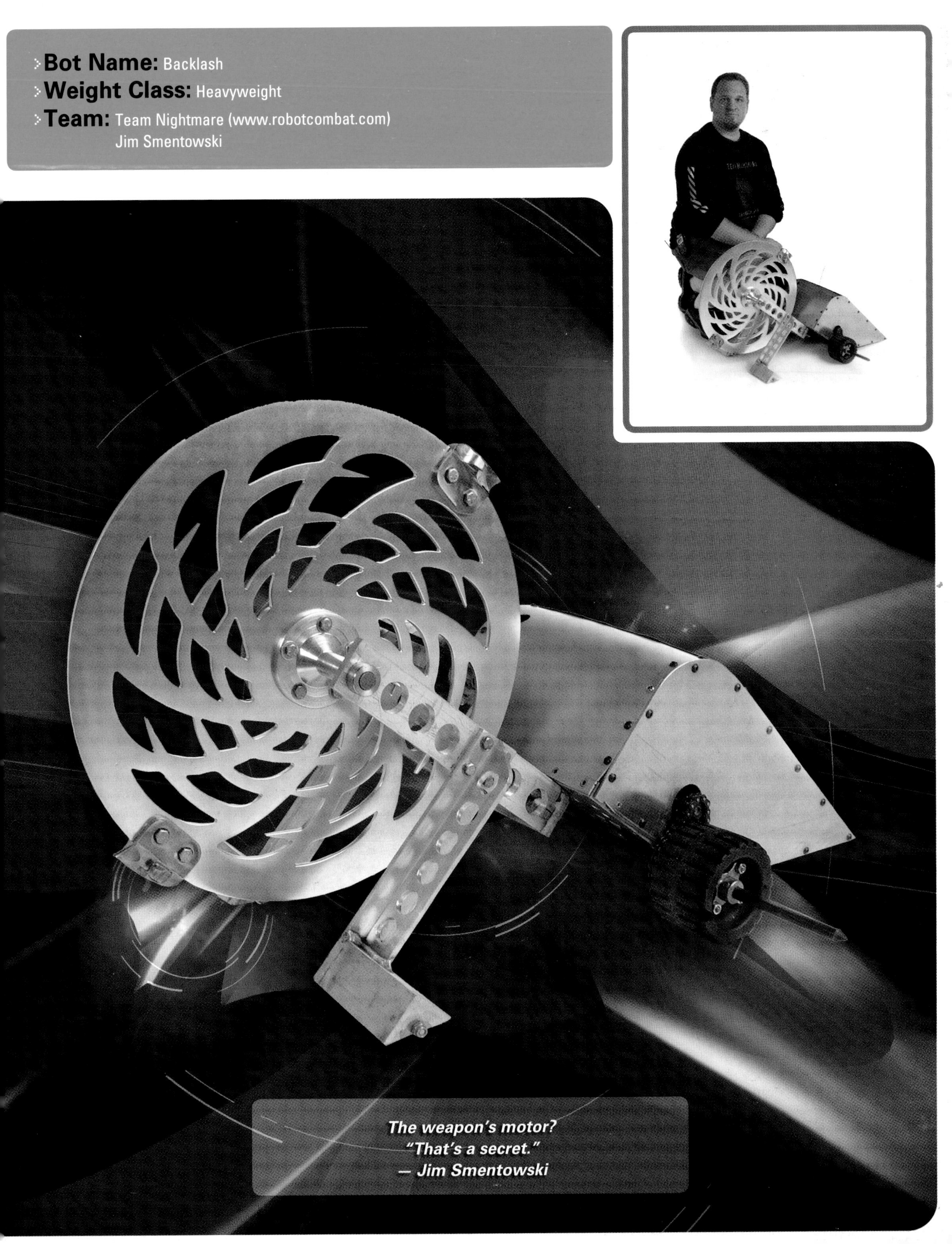

The weapon's motor?
"That's a secret."
— Jim Smentowski

BATTLERAT

"We have a gorgeous little shop, and we build weird, amazing stuff," says Ted Walters of his business, T.R.I.X. Rods and Racers in New Jersey. The vehicles that have rolled out of T.R.I.X. include Winston Cup (NASCAR) cars, the OnStar Batmobile, and the Heavyweight BattleBot BattleRat.

Battle Rat flips Tazbot

Racing Technology

"I've raced my whole life," says Walters. "I'm a crew chief, designer, builder, and fabricator. If I can build a car that can run at Daytona, I'm certainly capable of building a BattleBot."

With its two front fender humps, BattleRat resembles a miniature CanAm race car. Even Team BattleRat's crew dresses like a Winston Cup racing team. But BattleRat's resemblance to a race car is more than skin deep. "We're using all racing technologies in our BattleBot," says Walters.

Consider BattleRat's frame, made from 1-inch square steel tubing, and designed like a race car's rollcage. "It's very lightweight, and it's very strong," says Walters. "When you see a car at Daytona flip at 200 MPH, the driver just walks out. It's the same thing. You could hit my robot all day with a hammer and it won't do anything."

Navy-Aided Design

BattleRat was designed in about 15 minutes, says Walters. "I said, let's not get stupid. Let's not try to re-engineer the wheel. The wedge *does* work." For parts of the design that require a little more than pencil and paper sketches, Walters turned to the Naval Warfare Center, half a mile from his shop. "Those engineers and I get along very well," says Walters. "We go to the same place for lunch. I draw up the plans and then they tell me if it will work."

Walters got a little extracurricular engineering help with the design of BattleRat's chain-driven electric lifting arm. The arm can rotate through a full 360 degrees to lift opponents or extricate the Rat from stuck situations. "In the last Rumble," says Walters, "we used the arm to pick up another robot with a vertical spinner. That robot became my weapon, and we were dragging it around hitting [other robots]."

BattleRat is using new motors from National Power Chair, powered by NiMH batteries. The Rat's two drive wheels are chain driven. "We've never had a chain come off," he says. In fact, BattleRat has never had a major failure in the arena. "We've never been knocked out," says Walters. "We've finished every match. That's half the game."

Looking for the Win

"I approach BattleBots a little differently than most people," says Walters. "You can go there to have a nice time, or you can go there to win. And we're going to win." To that end, BattleRat's aluminum armor is being replaced with new, aerospace materials. "I won't tell you what materials we're using," he says. "It's lighter, it's better than titanium. It's awesome stuff.

"The materials we're using cannot be damaged. We'll make sure we have at least eight weeks of practicing and beating on the robot. We *will* win the next championship."

Bot Name: BattleRat
Weight Class: Heavyweight
Team: Team BattleRat (www.rodsandracers.com)
Mark Moses, Eric Moses, Sharon Caragias-Walters, David Baker, and Ted Walters
rodsandracers.com
BATTLE RAT
95.9
AIR graphix.com
CAUTION: DOES NOT PLAY WELL WITH OTHER ROBOTS!
CAPPADONA & SON
"You can go there to have a nice time, or you can go there to win. And we're going to win."
— Ted Walters

BIOHAZARD

Biohazard and Nightmare show no mercy

As of the printing of this book, Carlo Bertocchini's Biohazard is the winningest Heavyweight BattleBot of all time, with a very impressive 22 wins in only 25 bouts—more wins, in fact, than any other BattleBot in any division.

CAD to the Rescue

Biohazard's active weapon is an electric arm that slips under opponents and lifts them off the ground or tips them over. Bertocchini wanted that arm as big as possible, but he also wanted to keep Biohazard as low profile as possible. To achieve those goals, Bertocchini designed Biohazard in its entirety on a computer, using PTC's Pro/ENGINEER 3-D computer-aided design (CAD) software, before ever touching a tool to metal. Every single component of the robot has a corresponding model in Pro/ENGINEER.

Using CAD, Bertocchini was able to check component placement and fit, as well as accurately compute the bot's weight and center of gravity before starting to build. In fact, without 3-D CAD, says Bertocchini, "I wouldn't have even attempted it. I would have found a different hobby."

Amazingly, Bertocchini managed to stuff everything—drive train, lifting arm mechanism, armor, and all—into a package a mere three inches deep. Add an inch of ground clearance for the six wheels, and Biohazard stands only four inches tall. "It took quite a bit of design work," says Bertocchini. "Every detail of the robot was made more difficult by the constraint that it would fit into a three-inch package. The four-bar linkage of the arm mechanism was very complex. It could have been much simpler if I could have folded it down to, say, six inches instead of three."

Low-slung and Strong

But the advantages of such a low-slung BattleBot make it a worthy goal. With its very low center of gravity, Biohazard is almost impossible to flip over. It can deflect or dodge most weapons, which have to reach almost to the ground to hit it. And it can drive under opponents, using its entire body as a wedge.

Bertocchini's final goal was to give Biohazard maximum pushing power and maneuverability. Two-wheel-drive robots place some of the weight on castors, non-driven wheels, or part of the bot's body, reducing their pushing power. With four-wheel drive, the wheels drag at the corners when the bot makes tight turns. Biohazard's six-wheel-drive design places two extra wheels under the center of the robot. It can pivot on those center wheels, taking some of the pressure off the corner wheels.

An Engineer's Touch

Bertocchini is a mechanical designer by trade. His engineer's touch shows throughout Biohazard's design, from its amazing low profile to its intricately folded lifting arm; from the super-hard ceramic tiles protecting its underbelly, to the more than 700 machine screws holding the whole thing together. (Biohazard sports nary a weld.) Biohazard is beautiful on the outside, but Bertocchini's attention to detail runs deep. When he couldn't find drive motors small and powerful enough to suit him, Bertocchini bought the best he could find, took them apart, and then rewound the armatures to increase their power.

While the motors, and most of Biohazard's internal parts, have been replaced over the years, Biohazard's basic design and appearance have remained unchanged for a staggering six years—*lifetimes,* in the fast-evolving sport of BattleBots.

Bot Name: Biohazard
Weight Class: Heavyweight
Team: Team Biohazard (www.robotbooks.com)
David Andres and Carlo Bertocchini; not pictured, Carol Bertocchini and Rick Slagle

Biohazard is beautiful on the outside, but Bertocchini's attention to detail runs deep.

COMPLETE CONTROL

Complete Control lifts Subject to Change Without Reason

"BattleBots is a rock/paper/scissors competition," says Derek Young, "where you have wedges/lifters/spinners." With his latest BattleBot, Young wanted a robot that could beat all three. A ClampBot, he felt, had the best chance, *if* it could get hold of opponents without breaking. The resulting robot is called Complete Control. True to its name, Complete Control takes control of opposing robots, lifting them off the ground, carrying them around the arena, and feeding them to the arena weapons.

The Ultimate Wedge Killer

"Complete Control is the ultimate Wedge killer," says Young, although he admits that SpinBots can inflict serious damage to his robot's relatively delicate clamping and lifting arms. He's currently working on new construction and armoring techniques to improve his chances against them.

Complete Control's weapon is a two-part affair: an electric lifting fork to slide under opponents and lift, and a pneumatic grabber arm to grip opponents from above.

The forks are made of welded one-inch chromoly tubing. Dual electric motors deliver more than 1000 pounds of lifting force at the tips. Meanwhile, Complete Control's titanium grabber arm pushes down on the bad guy's top with roughly 60 pounds of pressure. Working together, they can grip an opponent like a thumb and fingers, and lift it completely off the ground—*voilà*, complete control.

Complete CAD

Twin 24-volt NiCad BattlePacks power the robot's six Astroflight Cobalt 40 electric motors—two for the lifting forks, and four for the drive train. (Astroflight motors are built for the RC model airplane industry.) Originally, an external 2:1 gear reduction from the gearbox drove the robot's rear wheels, which were chain-coupled to the front wheels. For Season 4.0, this gearing/chain setup has been replaced with a simpler "bow-tie" chain configuration driving both sets of wheels.

Complete Control's innards are bolted to a quarter-inch 2024 aluminum base plate, stiffened lengthwise with steel tubing to withstand the force of lifting other robots into the air. Aside from titanium panels front and rear, Complete Control relies on Lexan armor. "It is light, impact resistant, reasonably cheap, and easy to work with," says Young. "And it looks good; it can be painted from behind, so your paint won't scratch."

A Great Setup

By day, Young builds props for his local TV/movie industry. "It's a great setup," he says. "The job is flexible enough to allow me to devote whatever time I need to BattleBots, and I am also given free rein over the machines and materials."

In fact, says Young, during the last revision, Complete Control sat on his work table right beside a computer displaying a model of the robot in SolidWorks CAD software. "I would design a part in CAD, use CAM software to develop the tool paths, and then have the router table cut out the part, which would then be bolted onto the robot. This process was repeated until the robot was finished!"

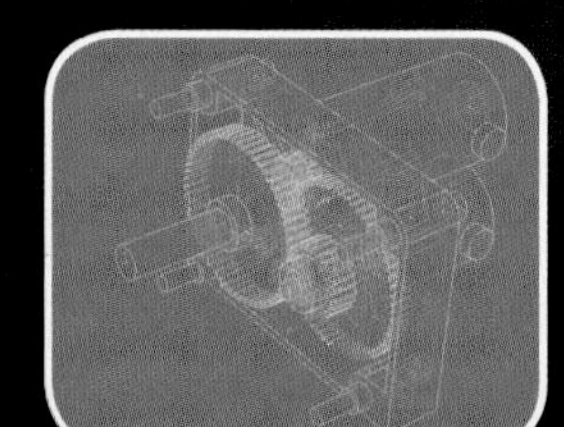

Partial design of the drive train

Unfortunately, Young's setup for last-minute changes at home is a bit less optimal. "My workbench is in the same room as my bed," he says. "My girlfriend wishes we had a garage (as do I), but she is thankful for earplugs."

Bot Name: Complete Control
Weight Class: Middleweight
Team: Automatum Technologies (www.automatum.com)
Derek Young
"BattleBots is a rock/paper/scissors competition, where you have wedges/lifters/spinners."
— Derek Young

DEADBLOW

Deadblow faces the Screw

Grant Imahara's BattleBot, Deadblow, is the number two ranked Middleweight BattleBot in the world, thanks in large part to its lightning-fast pneumatic arm.

Deadblow can use the spiked titanium hammer on that arm to mercilessly peck opponents into submission, Woody Woodpecker-style. (This little beauty has been known to deliver in excess of 100 blows to the head in a single, three-minute match.) Or it can catch bots on the backswing, lifting and tipping them from underneath, thanks to a pair of miniature forklift forks mounted on the end of the arm. Or maybe both, depending on what mood Imahara's in.

Deadblow's hammer is unlikely to penetrate the armor on better bots, but it can really rack up the points by delivering dozens and dozens of blows. Its super-fast speed allows it to strike an opponent repeatedly in the same place, making it all the more likely that something important will come loose on the inside.

A Prototypical BattleBot

Imahara prototyped Deadblow—twice—before he started building. "I just happened to be working in a place that deals with rapid prototyping and fabrication on a daily basis," says Imahara. That "place" is George Lucas's Industrial Light + Magic model shop, where Imahara spends his days as an animatronics engineer and model maker, and no one just "happens to be working there," despite Imahara's humble assertion. You think they let just anybody fumble around with R2-D2's innards? Not on your life, buddy! These guys are the best, and Imahara's no exception.

Working on a tight schedule (Imahara had six weeks to design and build the original Deadblow), he turned to the resources at hand. One of those resources was Autodesk's AutoCAD computer-aided design software, in use at the ILM model shop. Imahara used AutoCAD to create various computer models of Deadblow and finalize the design.

Another handy resource was a computer-controlled laser cutter, used to cut flat shapes out of transparent acrylic. Using scrap pieces of acrylic, the laser cutter, and his CAD designs, Imahara produced a see-through prototype of Deadblow. "The clear acrylic was really handy because I could see inside and underneath parts," says Imahara.

"For me," Imahara says, "nothing can replace having a physical model that you can walk around and turn over in your hand. Especially in this particular application, where the final product has to survive incredible amounts of punishment, having a physical model makes it possible for me to judge what might be difficult in CAD. I can imagine where an impact would happen and say, 'Okay, I'm going to need another support.'

"Plus, I like to make sure I haven't forgotten something before I fabricate parts in expensive aluminum or titanium."

Evolve or Die

BattleBot builders fall in love with a particular type of bot, says Imahara. "I prefer the hammer robot. It's very dynamic." But that doesn't mean that nothing has changed on Deadblow since it first rolled into the BattleBox in 1999. The most recent additions are the lifting forks on the arm and a pair of anti-wedge skirts on the sides. "I really like the design, and I've gone this far with it. It deserves to continue to evolve. If you left the robot the same every year people would figure you out. There'd be whole discussions on the Web on 'How to beat Deadblow.'"

Bot Name: Deadblow
Weight Class: Middleweight
Team: Team Deadblow (www.deadblow.net)
Grant Imahara
DEADBLOW
"For me, nothing can replace having a physical model that you can walk around and turn over in your hand."
— Grant Imahara

DEATH BY MONKEYS

Some bots are sentenced to death by Pulverizer,

Some are sentenced to death by Kill Saw,

but only the very worst are sentenced to... Death by Monkeys!

Death by Monkeys on the attack

Death by Monkeys

It's easy to get the impression that you need a machine shop and $40,000 to build a competitive BattleBot. Death by Monkeys is here to say that ain't so. Originally built for about $1,000, Death by Monkeys has competed in two events, making it to the semifinals in Season 4.0. And the robot's builder, Rob Farrow, started as the rankest of newbies.

"My training is as a psychologist," says Farrow. "I knew virtually nothing about electronics or engineering. But I saw BattleBots on TV and became hooked."

Keeping It Simple

"Death by Monkeys was my first bot so I decided to keep it *simple*!" says Farrow "I didn't want to mess with complicated pneumatics or active weapons on my first bot, and I think I made the right choice." Rather than designing the robot on a computer, Farrow laid everything out on a piece of wood, and then "went to town."

"Death by Monkeys is basically a sturdy metal box on wheels," he says, "with a thin titanium sheet on the bottom and top, and sharp pointy spikes out the front. I didn't know how to weld so I bolted everything together. Since I didn't have any machine tools, I spent some dollars on premade parts such as aluminum wheel hubs."

How long did it take to build? Longer than you'd think, says Farrow. "I had several false starts before I got the parts I needed and figured out how everything would work together. I also had very few tools: a measuring tape, a drill, and a hacksaw. Eventually I borrowed my father's jigsaw to speed up the metal cutting, but doing things without the right tools takes forever!"

Quick and Cheap

The robot uses NiMH batteries, driving four Ryobi 18-volt drill motors, one for each wheel. "Lots of people were using the DeWalt drill motors," says Farrow, "but the Ryobi motors have 90 percent of the power of the DeWalts, and are much cheaper and smaller." Despite its low-budget motors, Death by Monkeys is a very fast BattleBot, racing up to 15 MPH before ramming opponents.

Death by Monkeys' spikes are fixed at about 3/4" above the ground, too high to get underneath low Wedges, says Farrow, who's planning on adding an electric actuator to position the spikes up or down.

With upgrades and repairs, Death by Monkeys has cost Farrow around $2,000 so far, although not all of that money is in the final robot. "A lot of the cost of building these things comes from trying to solve reliability problems. Smash the bot into the wall several times and see what breaks. This means I spent a lot of money on things that didn't work, or didn't work well enough."

A Horrible Way to Die

But what about that name? "After spending thousands of hours studying primates in graduate school," says Farrow, "I developed a strong dislike of them." When it came time to name his BattleBot, Farrow wanted a scary, primate theme.

"Then one day it came to me," he says. "Being sentenced to death by monkeys would be the worst way to die."

Bot Name: Death by Monkeys

Weight Class: Lightweight

Team: Team Death by Monkeys

Pam Klein-Farrow and Rob Farrow; not pictured, Steve Judd and Nora Judd

"Death by Monkeys is basically a sturdy metal box on wheels."
— Rob Farrow

DIESECTOR

Dawn of Destruction takes a pounding from Diesector

"When I built Diesector," says Donald Hutson, "I built it from the jaws out." Hutson considered using a hydraulic "Jaws of Life" on Diesector, he says, "but it was too slow, so I ended up building my own.

"I played around with the jaw design in SolidWorks [3-D CAD software], but when it came time to do it, I just grabbed some steel, ran it through the bandsaw, and started making them. That's the best way to get it done."

Hard, Cold Jaws

Diesector's jaws are made from quarter-inch hardened, heat-treated steel, cut from a single length of plate. The jaw mechanism uses two custom-made linear actuators, with high performance motors and heavy duty shafts, rated at 1000 pounds each. "It's strong enough to destroy itself."

The jaws can touch the deck whether the robot is right side up or upside down. "That was quite complicated to achieve," says Hutson. "The jaws have to overbite themselves by 14 inches." To help control the jaws, and the rest of the robot, Hutson is using a programmable IFI control system onboard Diesector. "I can just push a button, and the jaws go to the deck."

If I Had Some Hammers

Diesector's hammers, Hutson insists, are also a very important feature. "The worst thing you can have happen is to have a 'dead side' on your robot. The hammers hammer in all four quadrants, whether the robot is right side up or upside down. In the last event, I actually used them to get myself off of another robot."

The twin ten-pound sledgehammers, machined from solid blocks of 6061 aluminum, are mounted to chromoly tubes, then rubber-coupled to the drive motors to absorb shock. "I'd much rather have a weapon fall off but save the motor," says Hutson. "The rubber coupling also gives them a nice follow-through."

Between driving Diesector around the arena and operating the jaws, Hutson was too busy to use the hammers successfully, so he called in a little help. "I give my roommates, David Cook and David Bleakley, each a hammer to control and that's all they do. It's very effective."

Lots and Lots of Motors

Diesector's two hammers, and its four wheels, are each driven by their own custom electric motor, made by Stature Electric. "With four separate drive motors," says Hutson, "you can rip a whole quarter of the robot out, and I've still got three working wheels."

But with six big motors and two heavy actuators, Diesector really sucks up the juice. "I use four gauge wire," says Hutson, "and it gets hot to the touch." Diesector uses the same high-current SVR sealed lead acid batteries as Hutson's Heavyweight BattleBot, Tazbot, but Hutson is considering switching Diesector over to NiCads: "Diesector has too many motors going."

Overall, though, Hutson is happy with Diesector's current configuration: "It has titanium armor, and a lot of offense. I can hit a robot at any corner, I can bite them, I can ram them, and I can drive them into the hazards.

"Of course, you can't make robots survive everybody. There will never be one perfect design."

Bot Name: Diesector
Weight Class: Super Heavyweight
Team: Team Mutant Robots (www.mutantrobots.com)
Mike Moore, David Cook, Dawn Bernstien, Donald Hutson, and David Bleakley
CINCINNATI
CINCINNATI
NPC
"You can't make robots survive everybody. There will never be one perfect design."
— Donald Hutson

DR. INFERNO JR.

Dr. Inferno Jr. vs. Bad Habit

Dr. Jason Dante Bardis came up with the name Dr. Inferno while a student at UCSB in Santa Barbara. His first name, Jason, is Greek for doctor, and his middle name, Dante, comes from Italian poet and critic Dante Alighieri, who wrote the Divine Comedy, which includes a trip through the Inferno. So... Dr. Inferno.

A Serious—But Cute—Robot

Bardis used Dr. Inferno as his radio DJ name for a time and built a combat-ready "plastic joke-bot," named Dr. Inferno as well. "After seeing how well the joke Dr. Inferno did," says Bardis, "I decided to make a 'serious' Dr. Inferno." A sequel, if you will: Dr. Inferno Jr.

"The joke," he says, "is that Jr. is bigger than Sr."

Dr. Inferno Jr. is intended to be a cute, amiable bot, "something kids could relate to," says Bardis. "I started with a toy plastic shell, put a strong skeleton inside it, and built a strong platform beneath it." The robot needed arms, so Bardis attached power tools: a circular saw and a reciprocating saw. Finally, he added four hinged flaps to the platform, acting as both offensive wedges and defensive skirts.

You might expect a Mechanical Engineering student to employ a little computer-aided design, but no, says Bardis, "It was a make-it-up-as-you-go design. I started with a pile of parts and then built the frame and drive train to go around them."

A Bot on a Budget

Bardis brought the first version of Dr. Inferno Jr. into the BattleBox for about $1,000. "I was on a tight budget," he says. "I used cheap, *cheap* power drills for the drive, batteries left over from a previous BattleBot, and a carbon fiber structure I made myself in my lab at UCSB. I bought the reciprocating saw arm from Sears with a Christmas present gift certificate."

The bot's final body used aluminum on the sides and carbon fiber/epoxy composite on the top, bottom, and skirts, with some extra spring steel to protect the Dr.'s bottom from frisky Kill Saws. "The carbon fiber material was *amazing,*" says Bardis, "stiff, strong, thin, light, and impact-resistant, but it's a pain to work with. It's dangerous to cut or drill—you need serious safety equipment to avoid inhaling carbon dust. Every time I handled it, I was guaranteed splinters and itchy rashes.

No More Itchy Rashes

"In rebuilding for Season 4, I was in a rush and had it up to here with carbon fiber, so I went with 3/8" Lexan. Spray-painted black, the Lexan looks really cool and worked great. The Lexan flaps survived both five-minute rumbles and not one bot got under the skirts. Plus, they self-sharpen as they drag across the BattleBox floor."

During the rebuild, Bardis also switched over to smaller, lighter NiMH batteries, saving about 10 pounds. He invested the weight savings in a beefier chassis and new weaponry: Dr. Inferno Jr. now sports dual circular saws with heavy saw blades.

The robot got new, stronger drive motors as well, but there was a wee problem: "The gears slipped off the motor shafts," says Bardis. For the next competition, he says, "I'm going to get steel gears and laser weld those suckers onto the motors so they'll never, ever slip. Ever."

Bot Name: Dr. Inferno Jr.
Weight Class: Lightweight
Team: The Infernolab (www.infernolab.com)
Dr. Jason Dante Bardis

"I used cheap, cheap power drills for the drive, batteries left over from a previous BattleBot, and a carbon fiber structure I made myself in my lab at UCSB."
— Jason Dante Bardis

EL DIABLO

Turtle Road Kill vs. El Diablo

When it comes to BattleBots, says Zach Bieber, "I like to make my designs aesthetically pleasing to the eye while keeping the aesthetics functional to the survival of the robot. I *hate* the box-with-wheels syndrome found so often in combat robots. These robots deserve to look cool and take on a personality."

Just Plain Cool

For example? "[For El Diablo,] I went with tracks because they are just plain cooler than wheels, and they make my evil robots look that much more maniacal. They also offer better traction than most wheels."

El Diablo's tank tracks, the monkey skull, the three-pronged spiked tail, the devil horns... they were all cool. But Bieber's robot is most renowned for its weapon: a 25-pound spinning barrel, studded with tool steel teeth and spun at 1500 to 2000 RPM. "I went with the spinning barrel for a weapon because no one had ever done it," says Bieber. "From a tactical and engineering standpoint, I chose to build it because kinetic energy weapons seemed to be very effective and, with a barrel, I could keep the mass of the weapon lower to the ground. This allows me to 1) attack the shorter robots and 2) make my robot invertible. The first El Diablo was built in my garage in about eight weeks. The first two to three weeks were design and the remaining five weeks were a feverish and obsessive attempt to construct the thing. I spent approximately $4,500 on the first Diablo.

"I started out with welded tube steel frames and graduated on to CNC'ed aluminum struts and framework (lighter, stiffer, tougher). I used aluminum plate for bottom armor (very resistive to the saws and prying lifters) and Lexan for top armor (more impact resistant to hammers and points)."

Highly Efficient

El Diablo's treads are driven by surplus Dunker high efficiency electric motors, with a planetary gear head. There are four motors mounted, each direct-driving a sprocket, which in turn drive the treads. El Diablo's spinning barrel is powered by one of the ever-popular Bosch 750s. The motors draw their juice from NiCad BattlePacks.

"I always over-volt my motors by 50 percent," says Bieber. "Most of my motors are 24 volt and I run them at 36 volts. This way I get higher performance, but don't risk spontaneous combustion before the match sees the 30-second mark."

Like all—or almost all—BattleBot builders, Bieber must build his creations on a budget. Many of the motors he's used come from surplus houses. But that doesn't really bother him too much. "I like working the way I do and under the constraints that I have," he says. "If it was made easier, I probably would slack off more and not live up to my own standard of quality."

Bot Name: El Diablo
Weight Class: Middleweight
Team: Team Diablo
Ron DeVaney, Heather Kyseth, and Zach Bieber
H & S ENTERPRISES, INC.
"I hate the box-with-wheels syndrome found so often in combat robots.
These robots deserve to look cool and take on a personality."
— Zach Bieber

ELECTRIC LUNCH

It's been said that breakfast is the most important meal.

Well, they got it totally wrong.

Pull your bot up to the table, sharpen your teeth

and then run like Hell!! Because you want none of this meal.

It's Lunch Time!

Odin II up against Electric Lunch

Super Heavyweight Electric Lunch is your basic, low-tech BattleBot, exemplifying Team K.I.S.S's design philosophy: "Keep it simple, Stupid." Steve Nelson and his father, Lowell Nelson, make up Team K.I.S.S. (and team S.L.A.M.). They became interested in robotic combat after attending a Robot Wars event in 1997. It was love at first sight, says Steve: "Before the event was over, we decided to build our first robot, S.L.A.M."

Keep it Simple, Stupid

Computer-aided design? Not for Team K.I.S.S. "Electric Lunch was designed on a bar napkin at my local watering hole," says Steve. "But we do use two types of CAD in our robot designs. Number one involves a cutting torch. We call it Creative Acetylene Design. We also build templates using cardboard. We call this Cardboard-Aided Design.

"Electric Lunch took about four months of three-day weekends, 16 hours a day, to build in the cold, snowbound garage at my dad's house, 75 miles away," says Steve. "If anyone ever wonders why I fight real mean, well, after freezing my bot off for three months I really want to kill something."

Don't Sit on That!

Electric Lunch originally sported a lifting arm powered by an electric winch motor. But a problem developed during stress testing, Steve says: "I decided to test it by jumping on the forks with my full body weight of 250 pounds, and I shattered the winch's motor shaft.

"I was out of money (as usual) but I had some quarter-inch plate steel lying around. So I built a bulldozer blade and a support frame from that. We bent the steel blade into shape using a large heating torch and an eight-pound sledgehammer. We did this at about 10:00 P.M. one night. It made the mountains ring like gunshots and sorta annoyed the neighbors."

Electric Lunch's base plate is made from a 3-foot by 3-foot piece of 1/4" T6 aluminum plate, reinforced with steel angle iron—some of it recycled from a bed frame—and steel pipe. The armor is 1/8" aluminum diamond plate, bent into a box shape.

The robot uses two 12-volt sealed lead acid (SLA) batteries, wired in series to produce 24 volts, powering two Bosch 750-watt motors rated at 1.2 HP each. Each motor connects to a jackshaft, which in turn drives both front and rear axles via a chain.

2.4 HP isn't much to drive a Super Heavyweight around the arena. In fact, says Steve, "Electric Lunch is one of the lowest-horsepower robots in its class, but I make up for the low power by using a very low gear ratio, trading speed for torque. To my surprise, robots with 12 times my power can't out-push me. That always makes me smile."

Heat Things Up

Some improvements are in store for Electric Lunch, says Steve. "It's getting a new frame and a weapon." If Steve's budget weren't so tight, Electric Lunch would also get titanium armor, bigger motors, and NiCad batteries. "Some insulation and a big heater for the garage would be nice also."

Bot Name: Electric Lunch
Weight Class: Super Heavyweight
Team: Team K.I.S.S.
Steve Nelson and Lowell Nelson
"We build templates using cardboard. We call this Cardboard-Aided Design."
— Steve Nelson

GENERAL GAU

If General Gau isn't the only BattleBot built as a college thesis and named after a chicken entrée, well, then it's certainly one of the very few. The members of Team Goosebearys took their team name from the most popular lunch truck serving the Massachusetts Institute of Technology (MIT) campus they attended, and their bot's name from Goosebearys' delicious General Gau's Chicken (great with the fried rice!).

Team member William Garcia has been crazy about robots since he competed in the F.I.R.S.T. robotics competition in high school, and his mom, Nola Garcia, commands the whimsically deadly BattleBot, Buddy Lee Stay in Your Seat. It was only a matter of time before he built a BattleBot of his own.

When friend and classmate Ahmed Elmouelhi approached Garcia and Dave Arguelles with the idea of building a BattleBot for a senior project and thesis, General Gau was born.

Super-Sticky Wheels and More

The team engaged in some furious brainstorming, under the guidance of MIT Mechanical Engineering professor Alexander Slocum, searching for that special, innovative twist that would make their bot unique. What did they come up with? Magnetic wheels that grip the metal BattleBox floor, providing General Gau with traction *way* out of proportion to its weight. "We spent a fair amount of time tweaking our wheel design to get the maximum amount of downward pull," says Garcia. In fact, the team's biggest expense was the neodymium rare earth magnets used in the magnetic wheels. In the end, General Gau's two wheels pull the bot down with almost 100 pounds of extra force.

The bot employs two DeWalt 12-volt right-angle drills and their accompanying NiCad batteries, straight from the hardware store, to power its drive train. Each wheel is attached directly to the output of its drill's gearbox. "The design was simple and provided plenty of torque and speed," says Garcia.

General Gau's motors are mounted between two plates of special lightweight "laminate": plywood, sandwiched between thin aluminum sheets. All of the bot's components attach to those two plates. The bot's transparent carapace is made from sheets of 3/16" thick polycarbonate, thermoformed into shape. Its nearly impenetrable plow was cut from a 90-degree bend of 3/8" steel pipe.

The General originally sported dual spinning chain flails with padlocks on the ends. But, says Garcia, "the flails were not that effective." So, for now at least, Team Goosebearys has dropped the weapons and is concentrating on improving the bot's armor and their own driving prowess.

A Successful Thesis

Dave Arguelles designed the entire robot prior to construction, using SolidWorks 3-D CAD software. The manufacturing process was a lot less of a headache, says Garcia. All told, the team spent about two and a half months designing and building the General—time that was pretty tough to come by while maintaining an MIT student's schedule, says Garcia. But they completed the bot in time, not only to compete in the May 2001 event, but to serve as Arguelles' senior thesis, as well.

All three Goosebearys graduated from MIT in May 2001.

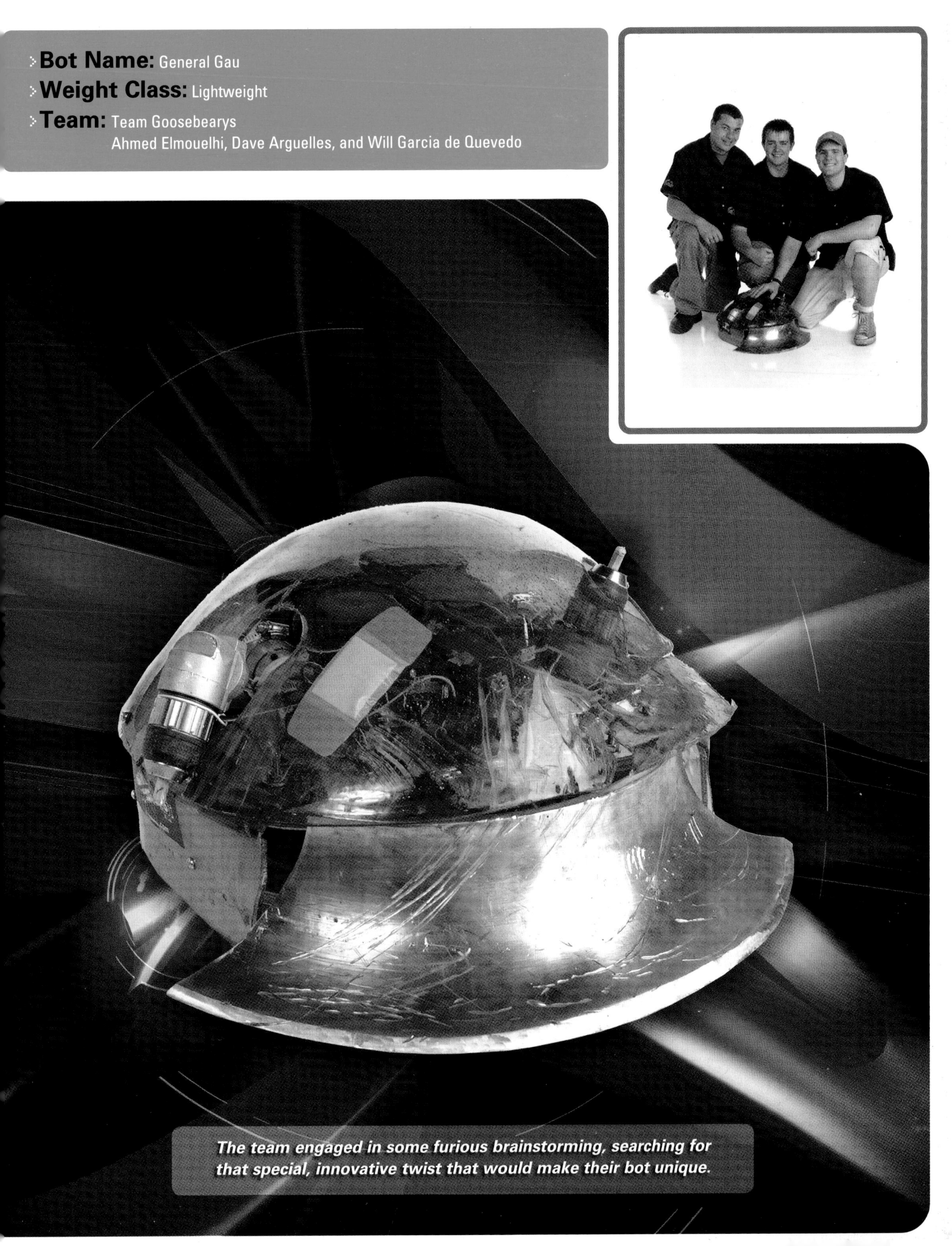

Bot Name: General Gau

Weight Class: Lightweight

Team: Team Goosebearys
Ahmed Elmouelhi, Dave Arguelles, and Will Garcia de Quevedo

The team engaged in some furious brainstorming, searching for that special, innovative twist that would make their bot unique.

HAZARD

"I entered Robot Wars 1996 on a whim and have been building bots ever since," says Tony Buchignani. "I have no formal training. In fact, I was a psychology major and then went to law school." In fact, Tony Buchignani is... *a lawyer*. But that's okay; it takes all kinds to build a BattleBots community.

Hazard takes on Heavy Metal Noise

Undefeated

As yet undefeated in battle, Buchignani's deceptively simple-looking robot, Hazard, is BattleBot's number-one ranked Middleweight (as of this writing). Its heavy, spinning blade both offends and defends, say Buchignani. "It's very difficult for an opponent to attack the robot without getting hit by the blade." Hazard also sports a small wedge on the front, as a backup in case the blade's not working.

Buchignani designed Hazard on the computer, using Rhino 3D and "a shareware program I downloaded from the Internet," he says. "I had machine shops build [the frame] from my drawings. I then installed all the internal and external components."

The robot's wheels attach directly to four DeWalt 18-volt electric drill motors. "The DeWalts are very sturdy," says Buchignani. "[They] have proved themselves in my robots since 1997. They come with gearboxes, weigh less than two pounds each, and are each capable of putting out 1 HP."

Hazard's deadly weapon is a 24-pound, 40-inch long blade of hardened tool steel, spinning in excess of 2000 RPM on a 1.5-inch axle of hardened chromoly. The blade is belt-driven by triple AstroFlight Cobalt 90 electric model airplane motors, delivering more than 1 HP each.

"I use NiCad [batteries] for the weapon motors, because they are best for sourcing the high currents drawn by the AstroFlights," says Buchignani, "and NiMH for the drive because they have the best run times at lower drive currents."

Stronger and Tougher

Buchignani put the first Hazard together in four months, for about $5,000. The robot has since been rebuilt twice, costing another $5,000 each time. "The batteries alone cost $1,200," he says.

Hazard has gotten stronger and tougher with every rebuild. The robot's current frame is made from half-inch 7075 aluminum bars. The top is armored in half-inch Lexan, the bottom in .09-inch 7075 aluminum. The frame and weapon box are made from half-inch 7075 aluminum.

Buchignani has also been able to reduce the overall height of the robot by more than two inches, bringing the blade down low enough to hit virtually every other robot.

For Hazard's next incarnation, the blade will be lowered another half inch. "I will be using custom motor mounts and wheel hubs, which will allow for very quick changes during the event. I also hope to change the top weapon bearing from a ball bearing to a tapered roller bearing, for increased reliability and strength."

So Middleweights beware: Hazard's coming back, and it's tougher than ever.

Bot Name: Hazard
Weight Class: Middleweight
Team: Team Hazard (www.legalword.com)
Alvin Buchignani, Tony Buchignani, and Adam Masri; not pictured, Sarah Buchignani

Buchignani's deceptively simple-looking robot, Hazard, is BattleBot's number-one ranked Middleweight.

HERR GEPOÜNDEN

Robot fighters everywhere
prepare to bow,
Herr Gepoünden's gonna make your bot
his Frau!

Herr Gepoünden vs. Death by Monkeys

Team Zwölfpack XII's name is faux-German for "Team 12-pack." Draw your own conclusions. In keeping with the team theme, they named their Lightweight BattleBot Herr Gepoünden. "It sounds better than its English near-equivalent, Mr. Beating," says Gepoünden's designer, David Otto.

Otto's background is in electrical engineering. "For my first robot," he says, "I was looking for a design that was fairly simple mechanically, and where I could apply my strengths in electronics."

Tornado Drive Engaged, Captain

Otto liked the idea of a two-wheeled ThwackBot but was wary of the ThwackBot's notorious disadvantage: they attack by spinning in a circle and, when they're spinning, they can't move around the arena. "I wondered," says Otto, "if it would be possible to make such a robot move linearly while spinning, by modulating the motor speed in sync with the rate of spin. Later I found that Blade Runner's builder, Ilya Polyokov, had envisioned a similar concept, which he coined 'Melty Brain.' I refer to it as 'Tornado Drive,' because I think it's more descriptive of the effect."

Herr Gepoünden took four to five months, and three to five thousand dollars, to put together. The robot is all aluminum with polycarbonate armor, and some strategically placed steel strips for Kill Saw protection. Otto used no computer-aided design for Herr Gepoünden's mechanical components. "I sort of designed it as I went along," he says. For the robot's special electronics, on the other hand, he employed Mentor Graphics Design Architect software for schematic capture and a free package from Express PCB for the circuit board layouts.

Administering a Poünding

The team lists an "unpleasant demeanor" as one of Herr G's weapons, but its primary punisher is its "wildly flailing heavy thingy"—an eight-pound sledgehammer head attached to the end of a titanium shaft. Otto is working on a modification to the hammer design, making the hammer head thinner and wider, so it will serve as a wedge as well as a bludgeoning device.

Herr Gepoünden uses NiCad BattlePacks powering two DeWalt 24-volt drill motors, held in place with limited edition Marcus-Mauler motor mounts built by the South Bay RoboWarriors (of Mauler 51-50 fame). The drill motor transmissions feed a chain reduction connected to ten-inch diameter wheels. The robot is controlled by Otto's patented "Spin Dr." control system, which, among other functions, implements the Tornado Drive.

Even More Pretty Lights

Herr Gepoünden has changed very little over time, says Otto: "The only significant change was the addition of the new 'Eye of the Tornado' rotational beaconing system." The beacon's job is to let the driver know which way the bot thinks is forward when it's spinning in place. The original beacon was implemented with a pair of laser pointers, but the lasers weren't bright enough to be effective. The new and improved Eye of the Tornado is implemented with a pair of RGB LED arrays, with a total of 192 LEDs in all. "So far," says Otto, "this system has proven effective at being very colorful and costing a lot of money."

Bot Name: Herr Gepoünden

Weight Class: Lightweight

Team: Team Zwölfpack XII

David Otto and John Karasawa; not pictured, Julian Otto, Alyson Otto, and Don Carroll

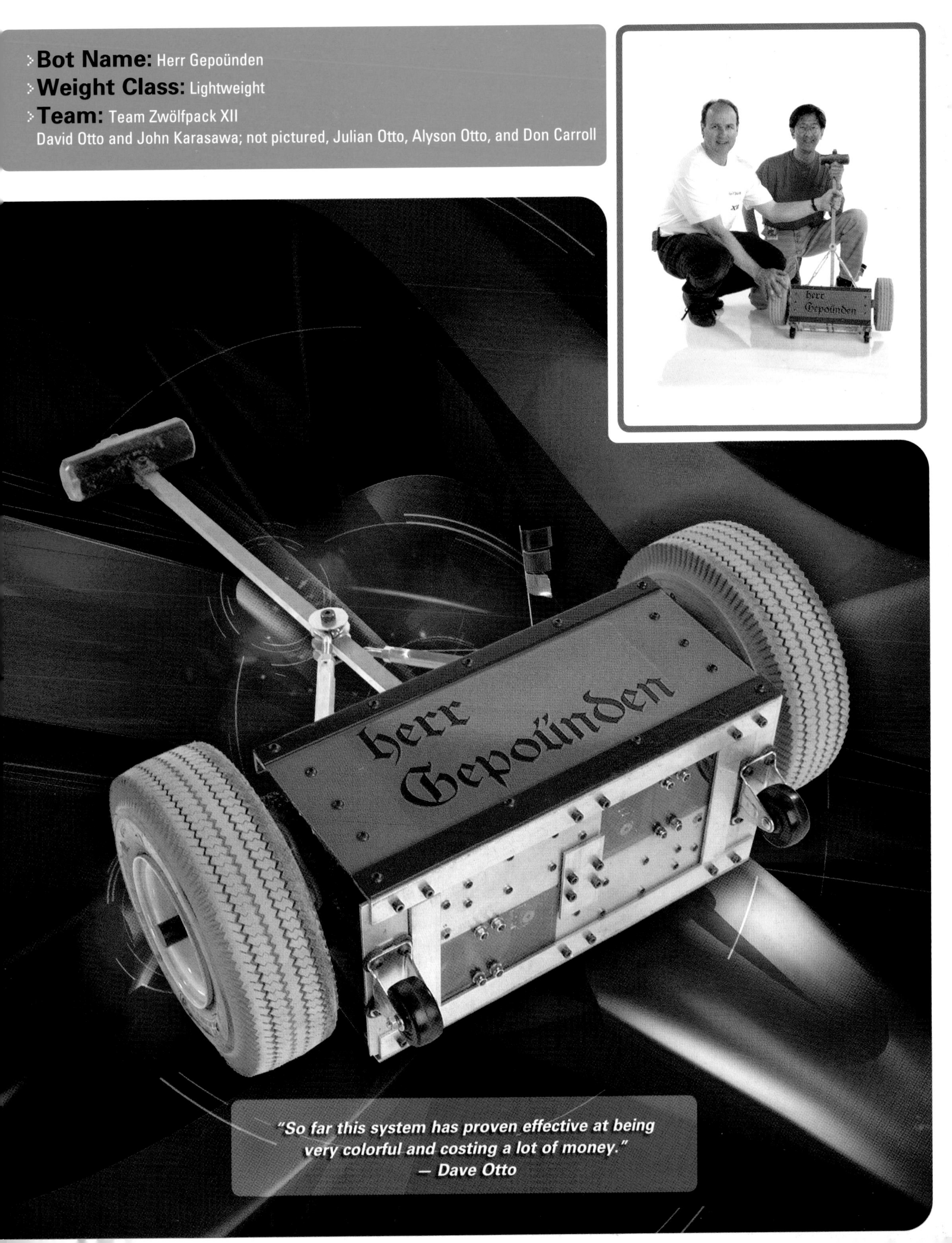

"So far this system has proven effective at being very colorful and costing a lot of money." — Dave Otto

HEXADECIMATOR

Hexadecimator battles Shark Byte

Team WhoopAss's Tim Paterson is well used to expensive hobbies. He and teammate Scott Ferguson have raced cars together in ProRally. "I spend ten to twenty thousand dollars per race on ProRally," he says. "My BattleBot almost pays for itself.

"I think my experience with ProRally had some positive influences on building a BattleBot," says Paterson. "ProRally is very hard on the car; one of my early mentors had the mantra, 'No pop rivets, no sheet metal screws.' I applied this kind of experience to the BattleBot, which deals with a similar high-shock environment."

Hexadecimator's design and construction spanned about five months and cost around $7,000, says Paterson, including capital equipment such as scuba tanks. Paterson did not use any CAD software in Hexadecimator's original design, but since then, the team has used LightWave 3D more and more to model their robots in the computer before building and rebuilding.

No Broken Welds

Hexadecimator's frame is built from quarter-inch thick 1-inch by 1-inch aluminum angle. "It's tricky to weld such thick aluminum," Paterson says. "I used a small MIG welder from Sears for Season 3, and the frame fell apart when it got hit hard. So I took a class in welding and bought a TIG welder the size of a small refrigerator. In Season 4 it still fell apart when it got hit hard (but the definition of hard went up). Goal for Season 5: no broken welds."

The robot's wheels are attached directly to beefy NPC-64038 electric wheelchair motors, powered by 12-volt sealed lead acid batteries. "We could potentially save weight using NiCad or NiMH [batteries]," Paterson says, "but we need weight in front, where the batteries are, for balance."

Hexy's weapon is a pneumatically activated flipping arm. High pressure air, delivered via a 150 PSI regulator from twin 110-cubic-inch paintball tanks, produces a lifting force of 600 pounds at the tip of the arm. The arm travels a full 90 degrees and can be used to right the robot, should it get flipped over in a match.

Puncture Resistance Testing

For Season 4.0, the team wanted tougher armor. They did extensive "puncture resistance testing," dropping a 40-pound spiked weight onto various materials to see how they held up. "Half-inch Lexan is the only thing of similar weight that beat eighth-inch titanium," says Paterson. "The titanium was penetrated by a 6-foot drop, while the Lexan held up to a 12-foot drop."

Hexadecimator now has a half-inch Lexan top, with eighth-inch titanium armor everywhere else. "Lexan is not suitable for sides or bottom because it is weak against saws, both Kill Saw and opponents' weapons," says Paterson.

Hexadecimator's new titanium skin set Team WhoopAss back nearly $3,500, says Paterson, but it works. "A trip over the Kill Saw in Season 3 resulted in our 3/16" aluminum baseplate being cut nearly all the way through," he says. "In Season 4, the titanium was just scuffed."

Bot Name: Hexadecimator

Weight Class: Heavyweight

Team: Team WhoopAss (www.teamwhoopass.com)
Scott Ferguson, Doug Franklin, Scott Franklin, and Tim Paterson

For Season 4.0, the team wanted tougher armor. They did extensive "puncture resistance testing," dropping a 40-pound spiked weight onto various materials to see how they held up.

HUGGY BEAR

The robot on the other side of the ring looks awfully scared.

I think somebody needs a hug.

And who better to give him one than ... Huggy Bear.

Huggy Bear and SABotage

Huggy Bear looks like an escaped prop from the set of Sesame Street's evil twin: a sort of menacing capital H with teeth. Five feet by four feet wide, and only five inches high, Huggy Bear is a ClampBot with a difference—it's not designed to crush opponents or even hold them up in the air. It just gives them a little *squeeze.*

A Most Singular Shape

How did this odd-looking BattleBot come about? "For months, everyone told me I had to see this show, BattleBots," says Huggy's designer, Dave Schultz. After finally seeing two episodes in January 2001, Schultz was sketching out ideas. He had a working BattleBot by that May.

Huggy's singular shape arose from a seemingly simple observation. "The arena hazards are much more powerful than anything I could build into the robot," says Schultz. "I figured I might as well use them." His idea was to build a robot that could latch onto opponents and deliver them to the arena weapons. "But I was frustrated that the Kill Saw kept throwing robots instead of cutting them," says Schultz. "I decided that I needed a way to get a good hold on the other guy, so I could cut him properly. Remember, you are supposed to hold your work nice and steady when you cut it."

A Nice Vise

In operation, Huggy mimics the vise he resembles, gripping opponents firmly on their sides and holding them "nice and steady." Because Huggy lifts an opponent's wheels off the ground, and presses opponents from the sides, he only needs a comparatively small force—about 80 pounds—to hold them in place. "That is more than enough," says Schultz, "because no robots have any mechanism that is designed to get them out of such a trap."

Huggy Bear originally used twin CO_2-powered pneumatic pressing arms to grab opponents. The latest version uses a single pressing arm for simplicity and weight savings. The front of Huggy's big, pontoon-shaped paws have been shaped into wedges to scoop up any bots too wide to fit Huggy's three-foot-plus kill zone.

Huggy Bear's wheels are connected directly to 18-volt DeWalt drill motors, driven by 18-volt Makita drill NiMH batteries, straight from Home Depot. "I didn't want to spend time figuring out the best motor," says Schultz, "so I went with what several others used. I call it cut-and-paste engineering."

Serendipity

Since finishing the robot, Schultz has discovered quite a few serendipitous aspects of Huggy Bear's unique design. "Huggy Bear is *super* easy to drive," he says. "I can easily place him where I want him. I thought I was a great driver, but it was really the design that made it easy to drive. It is easy to keep my nose and kill zone pointed directly at the enemy. And Huggy is so large, and so low, that he is nearly impossible to flip."

But best of all, says Schultz, "opponents get hung up on Huggy Bear with little effort from me." In fact, when Huggy got stuck in a Kill Saw crack during the Middleweight rumble in Season 4, three other robots got hung up on him.

Bot Name: Huggy Bear

Weight Class: Middleweight

Team: Team Huggy Bear — Donald Hammer, Tony Abejuro, Marie Schultz, Dave Schultz, Mark Schultz, Duane Osentoski; not pictured, Dave Kedrowski

Huggy Bear is a ClampBot with a difference – it's not designed to crush opponents or even hold them up in the air. It just gives them a little squeeze.

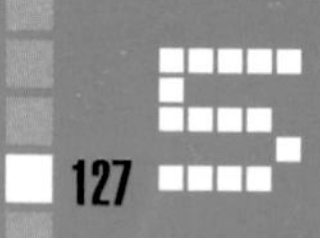

JAWS OF DEATH

Why waste time saving lives when you can be killing robots? It's the Jaws of Death!

Curt Meyers started building robots while attending Purdue University, where he met teammate and girlfriend Amy Sun. "I think my strange hobbies are what captured her attention," he says. He and Sun have been competing in robotic combat since 1996. Their newest creation is a unique, all-hydraulic bot-crusher named Jaws of Death.

Slow, Sure Death

Meyers has been interested in hydraulics, and their capacity for slow-but-sure robot destruction, for a while now, he says. Everything in the Jaws of Death, from the clamping jaws to its drive wheels, is hydraulically powered by Char-Lyn J2 series hydraulic motors. The motors are fairly heavy, but they are also fairly small and, being hydraulic, don't generate much heat.

With the system running at 2500 PSI, the robot's jaws exert about 15 tons at the tips! As the sharp tips pierce an opponent, it is drawn in closer. The jaws gape 4 feet wide and 24 inches deep when open, but close down to a scant 6 inches wide and 12 inches deep. The jaws close completely in about two seconds, and can grab most robots in less than a second.

Expensive and Loud

Hydraulics can be very expensive, says Meyers. "I really try hard to be resourceful. For example, a particular hydraulic system part cost $300. I knew that they only cost $150 in large quantities, but I only needed one." Meyers found a way: he purchased one part as a one-time "engineering prototype" at the $150 price.

The power source for Jaws of Death's hydraulics is a Briggs and Stratton Raptor engine, designed for go-carts, driving hydraulic motors. A 5 HP engine, it's been modified to produce 7 HP at 4,000 RPM, while running regular pump gas. "I rev it to full throttle and leave it there. The engine sounds great with its tuned CompCams exhaust pipe. It makes a great noise in the arena. Internal combustion engines are cool on a robot!"

The Right Design, the Right Materials

Meyers used PTC's Pro/ENGINEER CAD software to design the jaws and some other minor pieces of the robot. "Pro-E allowed me to adjust dimensions and leverage ratios to ensure that everything would work together," he says. "When the jaw was finished, the [model] was used to program a CNC plasma cutter."

Jaws of Death comprises a number of materials. "The materials are becoming as important as the design," says Meyers. The jaw supports are 7075 aluminum. The pins that mount the hydraulic cylinders and jaws are made from 2024 aluminum. The undercarriage and frame is 6061 aluminum. The jaws themselves are made of 4130 chromoly steel. The robot is surrounded by a custom roll cage of one-inch 4130 chromoly tubing. For the next version, the roll-cage will be covered by a piece of molded Lexan. Jaws of Death will also be getting a new *electric* part: a starter for the gasoline engine. "I should have [installed one] before our last fight," says Meyers. "We lost when our gas engine stalled."

Bot Name: Jaws of Death
Weight Class: Super Heavyweight
Team: Operation Boilermaker
Amy Sun and Curt Meyers
"The engine sounds great with its tuned CompCams exhaust pipe. It makes a great noise in the arena. Internal combustion engines are cool on a robot!"
— Curt Meyers

LITTLE DRUMMER BOY

He'll rip you, he'll flip you,
and beat you to a pulp.
He marches to the beat of a dangerous drum
... it's Little Drummer Boy!

Little Drummer Boy's designer, Steve Buescher, has always loved robots and design competitions. "When Comedy Central started airing BattleBots," he says, "the appeal was obvious. In March 2001, with encouragement from my wife, I finally decided to take a chance and build a robot."

Eraser vs. Little Drummer Boy

Uniquely Aggressive Destruction

After a season of armchair quarterbacking, Buescher had some specific goals in mind for his robot. "First, it had to be aggressive and destructive, so I wanted an active weapon. I wanted my robot to be invertible. Finally, I wanted a design that was at least somewhat unique.

"My solution was a vertically spinning drum. With the relatively smaller diameter, a drum design allows the robot to drive inverted."

Little Drummer Boy's frame is welded steel tubing. "A welded structure was the only choice for my team as we did not have access to a machine shop." Little Drummer Boy's first frame used locally available mild steel, since upgraded to 4130 Chromoly tubing.

New Armor and Motors

For armor, Little Drummer Boy used plastic UHMW (ultra high molecular weight polyethylene) panels, with a thin sheet of titanium on the bottom. UHMW is lighter and has better impact ratings than Lexan, says Buescher. "Turns out it was light, but also pretty useless. Standard impact strength tests aren't good at predicting usefulness for BattleBot armor." Little Drummer Boy now uses surplus ballistic armor panels from a local body armor manufacturer.

Little Drummer Boy first used four 24-volt DeWalt hammerdrill motors powered by NiCad batteries—two driving its four wheels, and two powering the drum, all chain drive. The robot has since been upgraded to four 18-volt Makita drill motors operated at 24-volts, driving timing belts rather than chains. "The small belts run much more efficiently at high RPMs," Buescher says. "And they weigh almost nothing, so I saved about a pound over the chain."

The robot's 6-inch diameter, 25-pound drum is made from a piece of steel tubing with two bars welded on for smacking the bad guys. The original drum spun at 3500 RPM, since increased to 5700 RPM.

Not Quite Ready

After two and a half frantic months of design and building, Little Drummer Boy was ready to compete in time for the May 2001 event. For travel, the robot was broken down, packed in four boxes and checked as baggage. "On the first day of check-in," recalls Buescher, "people were wheeling in assembled, ready-to-go bots while we were dragging in boxes of parts. We spent the first day reassembling only to discover, late that night, that part of the welded frame was damaged in shipping. We had to disassemble everything and wait for a friend with a welder to show up the next morning.

"Although I have a mechanical engineering degree," says Buescher, "I did not have *any* experience related to robots or machine design. I was not familiar with the motors, speed controllers, construction techniques, and materials used to build robots. I started from zero and learned what I needed to know as I went along.

"But my team was able to succeed with limited tools and knowledge. It is rewarding to compete at a high level despite being a 'garage hacker.'"

Bot Name: Little Drummer Boy
Weight Class: Middleweight
Team: Team Dangerous Drums (www.dangerousdrums.com) — Bill McGinnis, Kelli Buescher, and Steve Buescher; not pictured, Kory Gunnerson and Russ Barfield
"It is rewarding to compete at a high level despite being a 'garage hacker.'"
— Steve Buescher

LITTLE SISTER

This machine's a nightmare in metal

It's flipping Toss-tastic

Little Sister

Little Sister fights with Gammatron

"Our first robot was named Big Brother by my son, Joe," says Ian Watts. "He is the big brother to his two younger sisters Ellie and Megan. The next robot had to be Little Sister to even things up." Despite the name, big brother Joe, now nine, is Little Sister's preferred weapon operator. "Joe's very good," says Watts. "He goes when the weapon's engaged properly."

Although his first robot took six months, Watts put the current version of Little Sister together in only about four weeks. He describes his design process as "spending every hour of every day thinking about it." But he also uses AutoCAD 2D CAD software to help formalize the design before starting to build.

Like a Lamborghini, but Better

Before his recent retirement, Watts worked as an engineer for the BBC and as a lecturer in engineering at Brighton University. Until the robot bug bit him, he spent his spare time flying RC airplanes and building cars. His most recent project, a replica of a Lamborghini Countach, inspired Little Sister's structural design.

"I was thinking about the way the multibox space frame chassis works on the Countach," he says. "So Little Sister has a space frame chassis covered in skin. It's a very strong structure even though there's not very much metal in it. It also allows you to put everything into triangles; you can't deform a triangle."

Little Sister's weapon is a spiked flipping arm. The arm is fired by a CO_2 pneumatic actuator that produces 2,200 pounds of lift in about 100 milliseconds—1,100 pounds at the tip of the flipper. The robot's Bosch GPA 750 motors are powered by Hawker Genesis 17 amp-hour batteries, by way of a custom-made speed controller. Her wheels are chain driven.

The Strongest Armor

Little Sister's armor is actually a sandwich of 2-millimeter titanium, 32-millimeter polycarbonate, and 2-millimeter steel, over a frame of 20-millimeter mild steel box section, on top of an air gap. "Air's the strongest armor you can have," says Watts. "There's nothing to hit. And [they've] got to go that much further into it to do any damage."

Little Sister's titanium comes from a Russian Mig 29 fighter plane. "A friend of mine spotted [it] in a junkyard," says Watts, "and he nicked a bit of it for me. I guess it's flown all over Russia." Titanium's a bit hard to cut, so Watts acquired a plasma torch. He says. "I don't know if it'll work. It might just burst into flame."

Bot Name: Little Sister
Weight Class: Heavyweight
Team: Team Big Brother (www.teambigbro.co.uk)
Ellie Watts, Joe Watts, Cath Watts, Ian Watts, Megan Watts
Little Sister
"Air's the strongest armor you can have.
There's nothing to hit."
— Ian Watts

MECHADON

Mechadon creeps across the arena floor

Despite having competed in only four matches, Mark Setrakian's hulking 475-pound spider-bot, Mechadon, is probably the single most recognizable BattleBot in the brief history of the sport. Before creating Mechadon, Setrakian had competed in Robot Wars with his robots Snake and The Master. But for the first BattleBots in 1999, says Setrakian, "I wanted to start over again, to start fresh, to build something different from what had been seen in Robot Wars."

A Really Killer Puppet

Mechadon's design reflects Setrakian's day job, building and puppeteering mechanical creatures for films such as *Men in Black*. "It was more interesting to me to build a robot that was more creature-like," he says. "Mechadon allowed me to combine animal qualities and purely mechanical qualities. It has limbs like an animal, but it's also split into three sections that rotate independently, which nothing living can do.

"Mechadon is special to me in a lot of ways. I think of it as almost an art project. It was built very quickly—from nothing to a fully functioning robot in just four weeks. I don't recommend attempting such a thing. I was sleeping about one hour a day, so I think I poured a little bit of my subconscious into it."

Setrakian built Mechadon out of machined aluminum and hand-fabricated steel with a bare minimum of CAD assistance. Inside, Mechadon is an all-electric monster, comprising 14 high-voltage electric servo motors and actuators, powered at 170 volts by NiMH batteries. Mechadon has three rotating body segments, each with one pair of legs or claws, and its own receiver and electronics. The batteries, and the two large servos that control waist rotation, are in the center segment; slip ring assemblies transfer power to the outer segments. "It's a pain to make something pivot like that," admits Setrakian, "but the payoff is that it moves in such an interesting way. How something moves is just as important as how it looks, maybe more so. Looking at Mechadon in a still photo is only about 30 percent of the experience."

Making Mechadon move is itself a bit of a challenge, says Setrakian: "When I'm driving Mechadon, I'm literally puppeteering him. I had always intended to computer-control Mechadon, but I never got it mechanically stable enough that I didn't have to make constant adjustments."

The Coolest Thing I Could Build

Despite its 10.0 cool factor, Mechadon isn't the most effective BattleBot, in terms of its ability to win matches. But that's okay, says Setrakian: "At the time, I felt like designing to win was too restrictive. I wanted to branch out into walking robots, heavier robots... I wanted to push the envelope in every way. Although I did design Mechadon to have some threatening capability, I never really conceived of Mechadon as being a very competitive robot.

Mark Setrakian makes last-minute tweaks to Mechadon

"I tried to make the kind of robot that I imagined when people said 'fighting robots.' I built the coolest thing I could build." And cool it was, winning "Coolest Robot" and "Best Engineering" multiple times.

Says Setrakian, "Mechadon was my love letter to BattleBots."

Bot Name: Mechadon
Weight Class: Super Heavyweight
Team: Team Sinister (www.teamsinister.com) — Mark Setrakian; not pictured, Peter Abrahamson, Paul Rivera, Kyle Martin, Brian Morashita, and John Shea
"Mechadon was my love letter to BattleBots."
— Mark Setrakian

MECHAVORE

Weaned on aluminum...

Raised on titanium...

Straight from its post guarding the gates of Hell...

It's MechaVore!

Blade meets blade — Overkill vs. MechaVore

Robert Lawrence's Heavyweight BattleBot got its name one morning when, he says, "I sat up in bed and said the word, 'MechaVore,'" which has the following definition:

(mek' a vor), **n.** Any predominantly machine-eating robot of the order Mechavora, indigenous to the deserts of Arizona; mechavore = mecha (comb. form of mechi machine) + vore - vorous; loosely translated Latin term for a machine-eater. [Legend has it two of these guard the gates of Hell.]

Lawrence is the owner of Robert Lawrence Studio, in Tempe, Arizona. He creates a wide variety of one-of-a-kind sculptures, vases, mirrors, furniture, and door handles.

Lawrence eschews CAD for his robot design. "[MechaVore's] finalized design was conceived and executed in my mind before the first piece was cut," he says. "Time spent drawing pretty pictures is better used creating something real."

The Crowd Loves the Noise

MechaVore chews its opponents into bits with a heavy, horizontally spinning disk equipped with 5-pound, P-38 style hardened steel teeth. The teeth are designed to strike and rend, rather than cut. The disk spins at 3000 RPM, powered by a two-stroke, 6.5 HP gas engine, governed at 6000 RPM. Lawrence chose gas over electric for his weapon, he says, because the power-to-weight ratio is far superior. "Plus it's loud and nasty—the crowd loves the noise!"

MechaVore's disk is screaming death. Ask anyone at arena-side. In the Season 4.0 Heavyweight rumble, MechaVore was rear-ended and slammed into the BattleBox's Lexan sides. "Mechie chewed a nice gash through the Lexan and scared the hell out of cameraman and crew," says Lawrence.

The disk is low-slung, riding a few inches over the arena floor. "It puts the striking force in the kill zone on most robots, which is generally low to the ground and wide at the base," says Lawrence.

MechaVore is four-wheel drive for maximum traction while she chews on other bots. Her wheels are chain-driven 24-volt NPC electric motors, power supplied by Hawker Energy Cells. MechaVore's body is made of various thicknesses of aluminum, varying from 1/4" to 5/8". Her upper shell is made of 1/4-inch Kevlar composite, stylishly painted in classic MechaVore purple with dichroic flip-flop pearl. "It's lightweight, protects the engine, photographs well, and offers lots of real estate for sponsorship stickers," says Lawrence. (Hint, hint.)

Quest for the Nut

The first version of MechaVore cost about $3,500 and took a full six months to design and build. It was finished at the last minute, at the show. Since then, virtually every component has been replaced, except for the original aluminum frame. The robot's Wheel of Death gets deadlier every season. A new roll cage now protects the motor and allows Mechie to self-right automatically if flipped. The original side skirts and rear wedge are now gone. "They were unnecessary," he says, "and they showed too much damage for the judges."

In the quest for the Giant Nut, Lawrence bids you to remember that there are only three rules in BattleBots:

- Any bot can beat any other bot.
- You are only as good as your last fight.
- Losing really, really sucks.

Bot Name: MechaVore

Weight Class: Heavyweight

Team: Team Shrapnel (www.robertlawrencestudio.com/pages/mechavore/mechavore.html) — Ken Draper, Steve Bige, Dave Jelen, Robert Lawrence

MechaVore's disk is screaming death. Ask anyone at arena-side.

MINION

My robot can beat up your robot.

Minion is smokin'...

Super Heavyweight Minion, from Team Coolrobots, was BattleBots' first-ever Super Heavyweight champion, taking home the nut as the baddest bot in the newly created weight class in Las Vegas 1999, and then again in San Francisco 2000. Minion designer Christian Carlberg has these words of wisdom for aspiring bot builders out there: "Weapons don't win you a championship—it's all about your drive train."

Carlberg's background is in mechanical engineering. He's worked in aerospace, movie special effects, animatronics, and industrial robotics. He even helped create the original weapons for the BattleBox. When he sat down to create a Super Heavyweight BattleBot, says Carlberg, "I designed a robot that was all about drive train and strength... with a decent weapon." While some BattleBots rely on strategy and driving skill to win, says Carlberg, "Minion is all about power."

We're Number 1

After two first place wins and one second place, Minion was ranked number one going into Season 4. "Even so," Carlberg says, "we didn't want to sit on our asses. We had to keep improving it."

Fortunately, prior to Season 4, BattleBots increased the Super Heavyweight weight limit by 15 pounds, from 325 to 340. "For Season 3," says Carlberg, "we'd taken some of the power away from the drive train to improve the weapon. That extra 15 pounds made a huge difference for us. The weapon is as strong as ever, and we were able to get the power back into the drive train with better motors."

Minion now uses two 3 HP Lynch electric pancake motors, which produce higher torque and lower RPM. The batteries are sealed lead acid (SLA). Minion is a six-wheel-drive robot; the front and rear wheels are chain driven, while center wheels are direct-driven from their motors.

Unlike the electric motors, the 8 HP gasoline engine that powers Minion's weapon is unchanged, says Carlberg. "It's been so reliable and it's so powerful that there's no reason to replace it. But we have changed from a toothed blade to an inertial spinner blade: a thick disk with two whackers on it."

Armor? We Don't Need No Stinkin' Armor!

Despite being a really tough robot, Minion doesn't really have much armor, says Carlberg. "My attitude is that you want to put most of the weight into your drive systems, your batteries, and making the robot reliable. Make the robot really fast and put all the critical stuff deep down inside. *Then,* if you have weight left over, you put some armor on it." But Carlberg might be rethinking that strategy after Minion's loss in Season 4, when No Apologies punched a hole through an especially sensitive spot: Minion's radio receiver. "They had to hit a target no bigger than a half dollar," he says, "and they hit it perfectly."

Might I suggest a little more armor over that radio? "We already have a bit of titanium on the bottom," says Carlberg. "And you know for next season we're going to be putting a little bit of titanium in some other places!"

Bot Name: Minion
Weight Class: Super Heavyweight
Team: Team Coolrobots (www.coolrobots.com)
Luke Khanlian, Christian Carlberg, and Brian Roe
"Weapons don't win you a championship
—it's all about your drive train."
— Christian Carlberg

MOUSER MECHA-CATBOT

The contents of her litterbox are the nuts and bolts of dead robots,

she is the purrrfect fighting machine...

Mouser Mecha-Catbot!!!!!

Mouser Mecha-Catbot scratches Trilobot

Many BattleBots builders experience a little tension with their significant others, what with the time, money, and attention that a BattleBot soaks up. But not Team Catbot's Fon Davis. "It was her idea to begin with," he says. "We do it together." Davis and his wife and teammate, April Mousley, attended a robot fighting competition in 1996. Mousley was hooked on the idea of the couple building a combat robot of their own, but one with a twist. "April thought we should do a robot that was different than all the macho stuff we saw." The result was Mouser Mecha-Catbot, an angry pink kitty with a nasty, spiked tail. "We tried to achieve a balance between style and practicality," Davis says. Mouser is so stylish, she's even got her own music video.

A Cheap, Effective Kitty-Bot

The robot took about five months of weekends and evenings to bring to life. The team designed almost every aspect of Mouser using Autodesk's AutoCAD and Rhino 3D, programs Davis is familiar with from his work as a model-maker at George Lucas's Industrial Light + Magic. "They eliminate guesswork in designing the very tight layout inside Mouser's body," he says.

Mouser Version 1 was relatively cheap, says Davis. "We bought a lot of surplus and got some free stuff from friends, and kept the cost down around $750." Since then, though, Mouser has consumed a *bit* more money. "We've never had the guts to really add up how much it's been, total," he admits.

Mouser's body is an armored shell of vacuum-formed pink polycarbonate, over a base plate and ribs of machined aircraft aluminum. Eighteen-volt NiCad batteries power two Jensen 12-volt high-torque motors, one for each of Mouser's two high-traction rubber wheels. "We bought the motors surplus for only $8 apiece," says Davis. "They were inexpensive and performed great." The motors, each with its own speed control, feed custom-built reduction gear boxes, which in turn drive the wheels via chains.

Mouser packs three weapons: CO_2-powered pneumatic steel lifting forks, a CO_2-powered pneumatic sharpened steel tail spike, and an 18-volt electric circular saw, spinning at over 1000 RPM. "We wanted multiple weapons," Davis says. "We thought it would be more exciting."

Mouser's Nine Lives (So Far)

Mouser's been scrapping with other robots since 1997, and although she's still cute and pink, she's changed a bit over the years. "We change the look a little every season," says Davis. "Since our first design we have added the steel lifting forks, gearboxes, different wheels, different batteries, different armor, the spiked tail weapon, and much more."

What if money were not an object? What direction would Mouser take then? "We'd love to make Mouser more cat-like in shape, with real legs," says Davis. "We feel strongly that preserving interest in the sport will rely on making, and encouraging others to make, legged robots. People are going to get tired of our glorified R/C cars. Let's fight some real robots!"

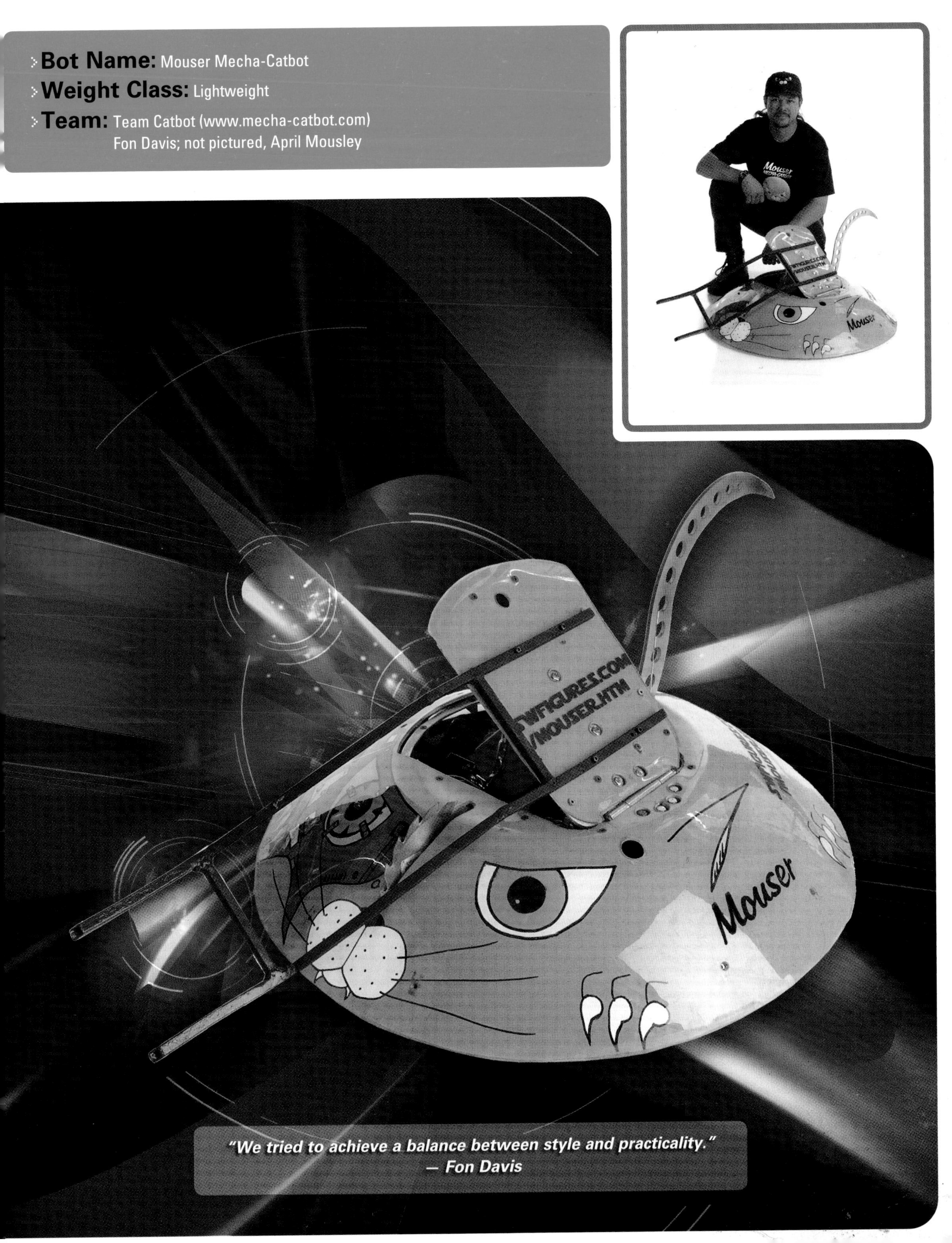

Bot Name: Mouser Mecha-Catbot

Weight Class: Lightweight

Team: Team Catbot (www.mecha-catbot.com)
Fon Davis; not pictured, April Mousley

"We tried to achieve a balance between style and practicality."
— Fon Davis

NIGHTMARE

In the BattleBox, Nightmare more than lives up to its name. Its giant toothed disk spins at nearly 1600 RPM (300 MPH) and disassembles other BattleBots, sending parts ricocheting off the BattleBox ceiling.

Biohazard vs. Nightmare

Spin It

Nightmare was the first of a whole new breed of weapon: the Vertical Spinner. Nightmare's designer, Jim Smentowski, says Nightmare's design was "inspired" by a vicious spinning robot called Blendo. "Actually," says Smentowski, "he kicked my butt.

"I wanted to have my own robot capable of dishing out that kind of damage in a very short time, but I wanted to do it in a different way. I found it was pretty effective if I took that rotating mass and turned it on its side, making an upward spinning weapon. Nobody had done that before."

And then, while digging around at a scrap metal yard, Smentowski came across some large aluminum disks. "They were just sitting back there gathering dust. So I grabbed them and turned one into the first disk that Nightmare ever used."

Dusty aluminum disks in hand, Smentowski tinkered the first incarnation of Nightmare together for about $3,500. "I like to go out and experiment with different ideas by putting things together, saying 'Let's see how fast this goes.'" In the end, Nightmare uses wheelchair motors to drive the wheels, while the weapon is powered by a 3.9 HP Magmotor, spinning at 4000 RPM, and geared down to spin the disk at 1600 RPM. Nightmare's weapon comes up to speed in two and a half seconds. After that ... look out.

These days, Nightmare's weapon disks are custom built, and made of a harder alloy than the original, but Smentowski still favors aluminum. Why? Spinners like Nightmare have to soak up as much damage as they dish out. When Nightmare slams a metal tooth into a 300-pound opponent's underside, it sends a tremendous shock through Nightmare's entire assembly. "Aluminum gives a little," says Smentowski. "A rigid steel disk wouldn't warp at all and it may even fracture." SpinBots such as Nightmare require lots of extra attention to those kinds of engineering details.

Tear Yourself Up

"It's funny," says Smentowski, "At the [May 2001] event, a lot of people brought vertical spinners that were completely self-destructive. They tried to copy the idea of Nightmare, and they didn't realize there was a lot more to it than throwing a disk between a couple of forks and spinning it up real fast."

But even Nightmare isn't entirely immune to self-injury. In one notorious bout against Slam Job in May 2001, Nightmare snapped its own motor mount—the robot equivalent of a broken ankle—when delivering its killer uppercut. Of course, that same uppercut sent the 200-pound Slam Job spinning through the air, chassis over teakettle, to land dead on its back, decisively ending the match and adding another KO to Nightmare's scorecard. If Nightmare lands one good punch, it rarely needs another.

Bot Name: Nightmare
Weight Class: Heavyweight
Team: Team Nightmare (www.robotcombat.com)
Jim Smentowski

Spinners like Nightmare have to soak up as much damage as they dish out.

PHERE

Out of the mist and darkness,
From the deepest corners of your mind,
Where dreams and reality intertwine,
Emerges the DRAGON of Destruction,
The spawn of a mechanical nightmare
And uncontrolled terror....
Phere!

Phere battles Anubis

"I have no background in robotics, says Team Xtremebots' Gaylan Douglas, "other than the small robot kits. But I have been watching BattleBots since the pay-per-view event and had dreamed of building one." When he found a sponsor willing to finance that dream, Gaylan wasn't going to let a little thing like no background stand in his way. Besides, wife and teammate Jan had lots of practical experience, says Gaylan: "Jan was raised on a chicken farm, so she is even handier at mechanical tinkering than I am. Give her a Dremel and she won't bother ya for at least a couple of hours!"

I Want It All!

Gaylan's goal for his first robot was nothing less than melding the best features of all successful BattleBots into one bot: "I wanted the stable drivability of a four-wheel-drive bot, the quick turning of a two-wheel-drive bot, an appendage to slap the other bot with, and the crazy unpredictability of a spinner.

"Jan and I also wanted a bot that had a personality," Gaylan says. He describes Phere's final look as a cross between an evil dragon and Darth Vader.

Gaylan used AutoCAD, a familiar tool from his "real" job in architectural design, to perform the major part of the design work on his computer. "Of course," he says, "there is always the garage banging that goes on to make the design work!"

Phere is built on a base plate design of 3/16" cold-rolled steel, with an interior framework of 1" steel tubing. Armor is more cold-rolled steel (what Steven and Lowell Nelson of Team K.I.S.S. call farm steel, because it's cheap, tough, easy to weld, and universally available).

Phere took almost a year to build and cost "considerably more than our four kids' college fund has in it right now!" says Gaylan. "More than the hard cash are the blood, sweat, and tears equity involved in a BattleBot."

Pumping Up in the Off Season

Like most BattleBots in the off season, Phere is getting several updates. "The only constant," says Gaylan, "is the Sinister Dragon attack wedge, and the 36-inch spun steel Dome of Destruction with corkscrew blades."

Phere's 36-inch spinning dome is being upgraded with twin 2.9 HP electric Magmotors, which should spin the 82-pound dome, donut blades and all, to nearly 200 MPH. Phere's drive train is getting new, more powerful NPC 84088 wheelchair motors, providing 2.5 HP to each of Phere's four wheels. The motors will all be powered by new NiCad BattlePacks, pricier but lighter than Phere's old SLA batteries.

If I might suggest a revision? Phere's new Donut Blade, added to the bottom of the Dome of Destruction, proved the bot's downfall in November. In its first fight, a hit to the dome caused Phere's untested blade to chop its own kill switches off, rendering the bot dead on its feet. "We were just a 336-pound toy to be shoved under Pete's hammers," laments Gaylan. I expect a bit of a redesign on this feature before the next match.

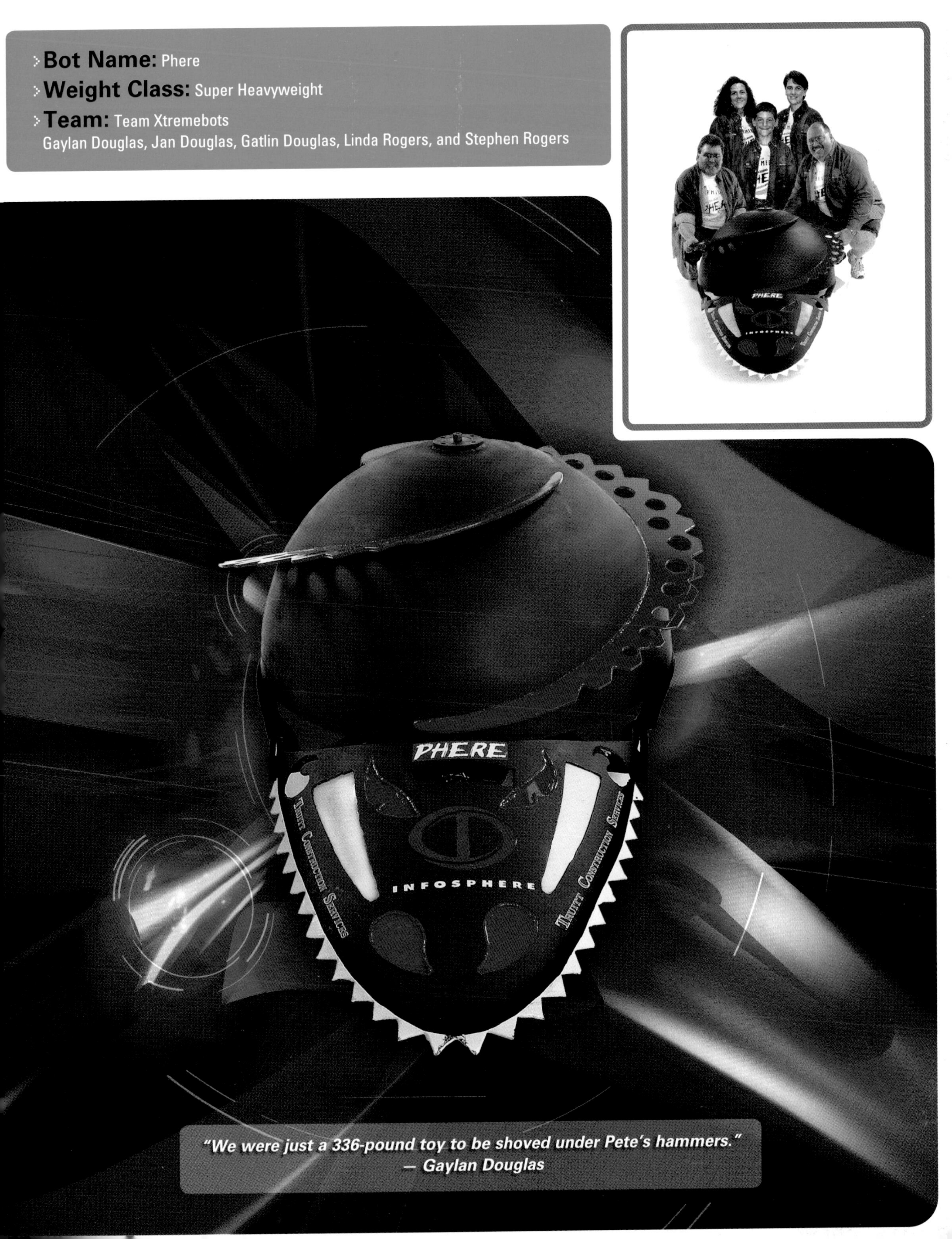

Bot Name: Phere

Weight Class: Super Heavyweight

Team: Team Xtremebots
Gaylan Douglas, Jan Douglas, Gatlin Douglas, Linda Rogers, and Stephen Rogers

"We were just a 336-pound toy to be shoved under Pete's hammers."
— Gaylan Douglas

RONIN

Ronin in full glory

"I'm really interested in Japan," says Ronin's creator Peter Abrahamson, "and I have a huge affinity for Japanese samurai movies." As if we couldn't tell by the three huge, red Japanese battle flags that his bot carries into battle.

A *ronin* is a masterless samurai, working as a sword for hire, the eastern equivalent of a freelancer. As a freelancer himself—by day, Abrahamson assumes the role of a mild-mannered special effects artist—Abrahamson liked the metaphor. He named his company Ronin Special Effects.

The Tank from Burbank

When it came time to name his BattleBot, well, what better image than that of the lone samurai? Heck, the original Ronin even wielded a sword.

That sword was soon replaced, first by a gasoline engine and a vertical circular saw blade, and more recently by a heavy, horizontally spinning disk driven by surplus electric treadmill motors. For the next event, Abrahamson is modifying the weapon arm to allow the disk to spin either horizontally or vertically.

Love Them Rhomboids

But some things will never change, says Abrahamson, especially Ronin's defining, rhomboid "tread pods." "I've always loved treads," says Abrahamson. "Tanks are beautiful, amazing pieces of machinery. There's one funky World War I British tank called the Mk. IV. The treads went all the way around. I loved the look of it.

"Scott LaValley also had a great little parallelogram-shaped tank, called DooLittle, which blew me away. That also influenced Ronin's design."

Ronin is like a pair of treads with no tank; there's no body, per se. Rather, Ronin's two tread pods are connected via a length of two-inch diameter metal tubing. The tread pods can pivot, separately, swiveling up to 90 degrees apart. "I can climb over bizarrely shaped robots without getting high-centered," says Abrahamson.

That connecting tube design, borrowed from teammate Mark Setrakian's robot, The Master, also allows Abrahamson to clamp on any number of interchangeable weapons—or at least such number as he can find the time and money to build.

High Voltage

While most BattleBots run on 18- to 24-volt current, Team Sinister's robot runs at 144 volts. "We run much higher voltage at much lower amps," says Abrahamson, "because it's more efficient." Ronin also uses Nickel Metal-Hydride (NiMH) batteries, rather than more common sealed lead acid (SLA) or NiCads. Taken together, Ronin's electrics give it sufficient juice for two full fights.

Each tread pod holds one over-driven Baldor servo motor, controlled via an Advanced Motion Control servo driver card—equipment familiar to Abrahamson through his animatronics special effects work.

The motors feed through a right-angle planetary gear head into a custom transmission mounted inboard of the pod, and finally to the front drive sprocket. The transmission uses belts rather than chains, says Abrahamson. "A belt can slip if it has to and save my system. I'd rather have a little less torque than a broken robot.

"But whatever you build, it's never enough. I can't believe what a devastating environment the BattleBox is—what a test bed!"

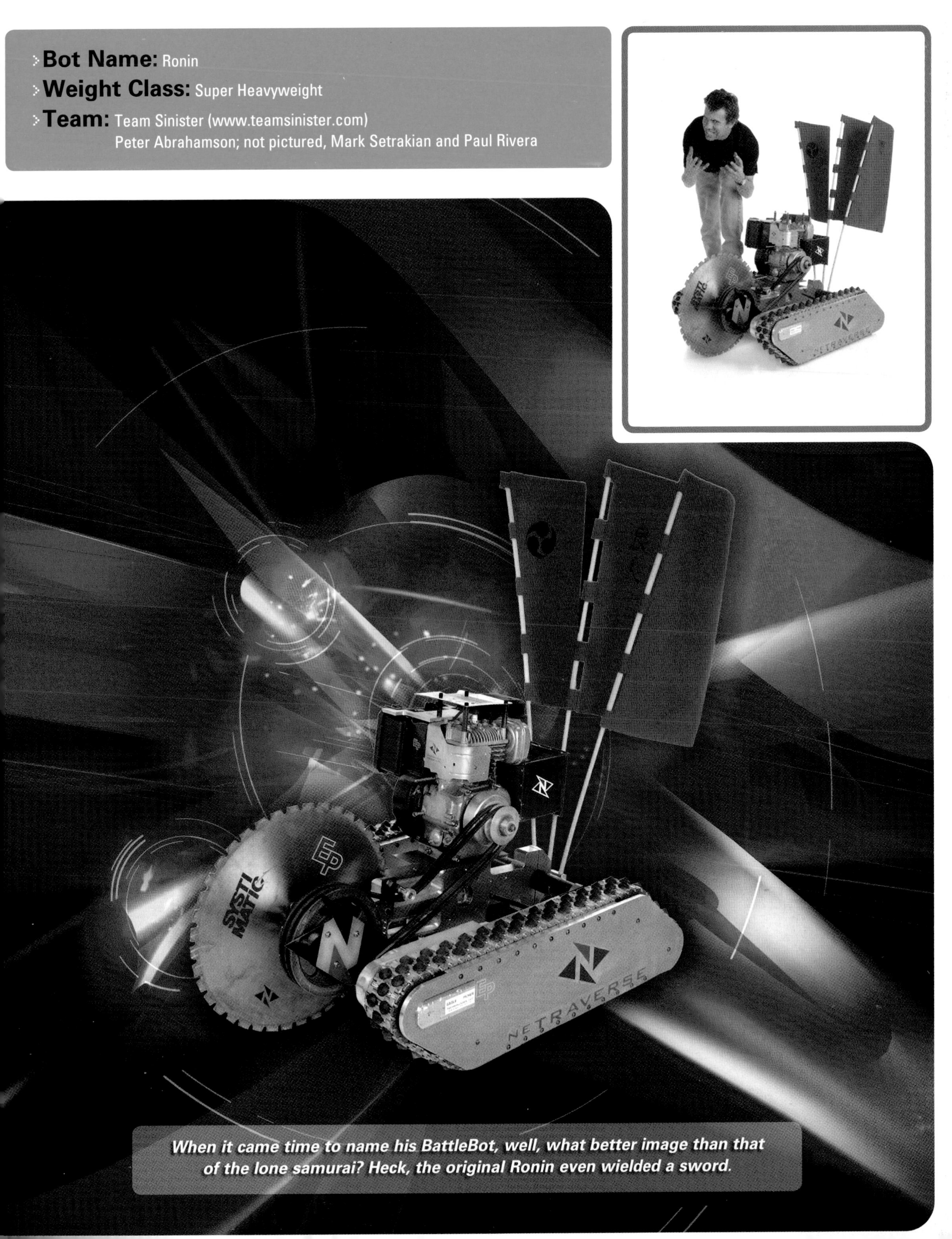

Bot Name: Ronin
Weight Class: Super Heavyweight
Team: Team Sinister (www.teamsinister.com)
Peter Abrahamson; not pictured, Mark Setrakian and Paul Rivera

When it came time to name his BattleBot, well, what better image than that of the lone samurai? Heck, the original Ronin even wielded a sword.

SALLAD

Sallad fights Fang

Dallas Goecker built his Lightweight BattleBot Sallad ("Dallas" spelled backward) on a budget sure to inspire garage tinkerers everywhere. It took him a scant four weeks to put the first version of his robot together in his garage using a hacksaw, table sander, drill press, and tap set. The initial price tag was only about $1,000, he says. "Half of my total cost was in a new radio transmitter and receiver. I made or customized everything to keep the costs down. The gear boxes were $2 each, and I customized them to change the gear ratio and accept a larger motor. The motors were $5 surplus drill motors. They advertise the motors as being 12-volt motors, but I think they are really designed for 9.6 volts. I am running them at 14.4 volts to get more speed.

"I made my own speed controller. The wheels were aluminum disks with conveyor belting wrapped around them for traction. And the aluminum was all from scrap yards."

Sallad is built around a 1/4" aluminum base plate with two 3" by 3" pieces of 3/16" aluminum angle for the wheels, motors, and gearboxes to mount to. Sallad's top is covered in 0.092" thick sheet metal, and the robot is surrounded by four 3" by 1 1/2" 6061 aluminum "U" channel bumpers.

Getting Tossed by Sallad

Sallad's weapon is a lifting arm, made from a single piece of 2" by 2", 1/4" thick square aluminum tubing. The arm can swing 180 degrees up and down and can be used to lift, scoop, ram, or pound on robots, or even to set Sallad back on its wheels should it get ... ahem ... tossed.

The current arm is the fourth-generation design. "I completely machined my own custom gearbox with a worm drive system and clutch," says Goecker. "The gearbox worked great. The clutch worked great. This last season I added a cam system to the arm that increased the lifting torque of the arm by a factor of ten. I now have 200 pounds of lifting force through the first six inches of lift." Although he originally designed Sallad without benefit of CAD, Goecker now uses PTC's Pro/ENGINEER software for all his design work.

Sallad's drive and weapon are powered by "cheap" 1500mAH NiCad battery packs. "You pay three times as much for 'high performance' NiCads," Goecker says, "but you do not get 2500 milliamp hours out of them [in a three-minute match]."

Reliability Issues

Currently, Sallad's biggest problem is its reliability. "It's because of the cheap components I use," says Goecker. "It has a lot of redundancy, which usually keeps it running, [but] after most matches I have to replace something that broke or burned."

Not to worry. He has a whole new Sallad designed, with an awesome new drive system and a powerful new arm—as soon as time and funding become available.

"If money was not an object," says Dallas, "I would quit my job and play in the garage the rest of my life."

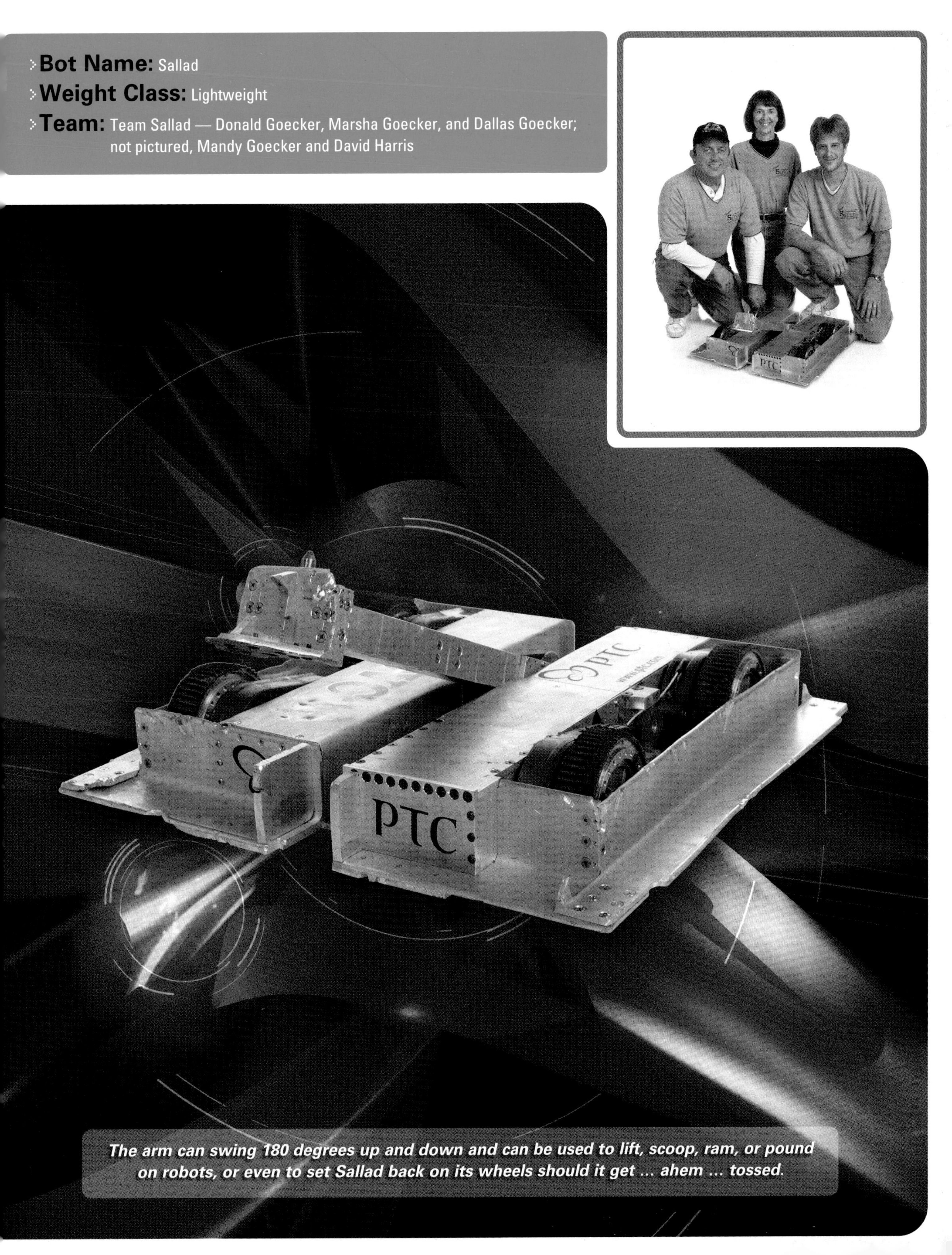

Bot Name: Sallad

Weight Class: Lightweight

Team: Team Sallad — Donald Goecker, Marsha Goecker, and Dallas Goecker; not pictured, Mandy Goecker and David Harris

The arm can swing 180 degrees up and down and can be used to lift, scoop, ram, or pound on robots, or even to set Sallad back on its wheels should it get ... ahem ... tossed.

SHARK BYTE

Come on in, the water's fine, trust me...

Shark Byte dances with Hexadecimator

Arndt Anderson didn't come to BattleBots looking for money or fame, he says. "I just wanted to destroy stuff in the arena. I have been into destruction all my life. BattleBot just made it legal." Of course his BattleBot, Shark Byte, had to wield a spinning weapon. "Win or lose," he says, "it offers the most potential for massive impacts, hence more parts to go flying."

Shark Byte's weapon is a spinning top plate made of 3/4" 7075 aluminum, with heat-treated stainless steel blades sticking out in the four cardinal directions. The weapon is 47 inches across, weighs in at 65 pounds, and spins at up to 1750 RPM.

It's the Unique Touches

What's unique about Shark Byte's spinner, says Anderson, and what makes it so strong, is the way he spins it: "Most [SpinBots] spin an axle with the weapon attached to it. Mine is a free-wheeling axle with wheel bearings out of a car. I drive the weapon with a rubber wheel underneath that spins against the bottom of the disk. That does two things: first, the axle point is rigid as hell; it's not going anywhere. Second, the rubber wheel rubbing against the bottom of the plate acts like a slip clutch when [the weapon] hits something. I'm not going to break motor parts."

Anderson's robot is unique in other ways, too. Where most BattleBots bolt components to a bottom plate, Shark Byte has a 3/8" magnesium top plate. The robot's components—the motors, the batteries, the wheel structures, the aluminum and titanium armor, and so forth—are bolted upside down to this one plate. "It just worked out that way," Anderson says.

And then there are the tiny wheels, only three inches in diameter. "I use really, really small wheels," says Anderson. "I have a theory that, the smaller the wheel, the better traction you get on the arena floor." Shark Byte is a four-wheel-drive robot, with one drive motor per side. Each motor drives a reduction gear box attached to a front wheel. The back wheels are driven off of the front wheels by timing belts.

"I'm a mechanical guy," says Anderson, "and I went to my first event with the machine well built mechanically. Part of BattleBots is building something that's going to last three minutes. In Season 4.0 we proved to everybody how stout we were. We were getting pounded by the Pulverizer, but we were spinning even while it was hitting us. We broke the arena and put a big hole in the Lexan wall." No one got hurt. Pete fixed the hole immediately.

Electrical Troubles

Shark Byte is mechanically robust, but it's had some problems. "I'm still fighting with the electrical," Anderson admits. "I was using Leeson 24-volt motors. We were pumping 36 volts into them, and we fried the armature brushes. We're switching to Briggs & Stratton E-Tek motors, because of their better horsepower rating and their low profile. We are also switching to NiCads, because they dump their juice quicker in a three-minute fight.

"So I've got the right motors and the right batteries coming. Hopefully the robot can perform the way it's supposed to perform."

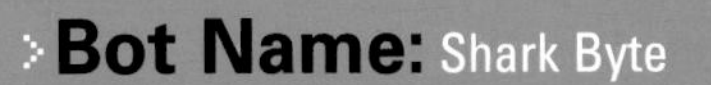

Bot Name: Shark Byte
Weight Class: Heavyweight
Team: Team Shark Byte
Cody Rose and Arndt Anderson

"Part of BattleBots is building something that's going to last three minutes. In Season 4.0 we proved to everybody how stout we were."
— Arndt Anderson

> SNOWFLAKE

She's rough, she's tough,

she'll beat you sure enough...

It's SnowFlake!

"My father was building robots," says 13-year-old Jacey Ross, "and asked if I wanted to have my own to drive." Well, of course she did! Her robot is the Lightweight SnowFlake.

Snowflake in the Rumble

A Family Affair

Team Toad is a family affair, and everyone in the family has at least one combat robot to call their own. Most sport wintry names such as Iceberg, Windchill, and Ice Cube. "Since a Lightweight is the smallest class," says Jacey's dad, Michael "Fuzzy" Mauldin, "we figured a snowflake is the smallest icy object." SnowFlake was designed as a two-wheeled "mini-me" edition of Dad's four-wheeled robot, FrostBite 2.0, using the same motors, wheels, and type of weapon.

SnowFlake's weapon is an unpowered snowplow, three pounds of quarter-inch thick 5052 aluminum. SnowFlake's body is made from more bendable, eighth-inch thick 5052 aluminum. Dad designed SnowFlake's body and blade in CorelDRAW, and then had them bent and welded at a local fabrication shop.

SnowFlake's two wheels are mounted directly to a pair of NPC 60522 right-angle wheelchair motors, powered by lightweight Panasonic NiMH racing packs. SnowFlake's motors originally ran at 24 volts. For Season 4.0, BattleBots raised the weight limits for Lightweights from 58 pounds to 60 pounds, and the team was just able to squeeze in a beefier speed controller and sufficient additional batteries to run the motors at 36 volts. The new, improved SnowFlake is now 43 percent faster and, at 59.5 pounds, just barely squeaks in under the Lightweight weight limit.

Good Mojo

"The particular [speed controller] we used was the same one we loaned to Carlo for Biohazard last season," says Dad. "We figured a little 'Biohazard mojo' would be a big help." That mojo got SnowFlake into the TV rounds, where she faced Jonathon Ridder's super-destructive Lightweight, Ziggo.

"Ziggo's first hit nearly ripped the plow blade completely off [our] little bot," says Dad, "and SnowFlake's master switch failed, disabling it. Just before SnowFlake was counted out, Ziggo hit it again, and the master switch miraculously turned back on. SnowFlake was moving under its own power again. But the second chance didn't last long—another hit from Ziggo, and SnowFlake was dead a second time. We yelled over to Jonathan Ridder, 'Hit it again!' but this time SnowFlake was dead for good."

Fortunately, SnowFlake had two spare bodies and was rebuilt and looking good in time for Jacey's television interview a month later.

Changes for next season? "We plan to make the master switch more robust," says Dad.

Bot Name: SnowFlake

Weight Class: Lightweight

Team: Team Toad (http://www.lazytoad.com/teamtoad/snowflake.html)

Jacey Ross; not pictured, Kelsey Ross, Danny Mauldin, Michael "Fuzzy" Mauldin, and Debby Mauldin

SnowFlake was designed as a two-wheeled "mini-me" edition of Dad's four-wheeled robot, FrostBite 2.0, using the same motors, wheels, and type of weapon.

SON OF WHYACHI

Son of Whyachi contends with Hexadecimator

Son of Whyachi is the robot that changed the rules on walkers. Terry Ewert designed the robot to take advantage of a well-known but little exploited clause in the BattleBots technical regulations: walking robots were allowed a 50 percent weight bonus. While wheeled Heavyweights were limited to 210 pounds, a walking robot could weigh as much as 315 pounds. The trick was to come up with an effective walking mechanism that didn't add much weight to the design.

Ewert invented a unique, shuffling drive train that qualified the robot as a walker, while avoiding drawbacks of a traditional walking robot: high weight, high center of gravity, and slow speed. He then threw lots of extra weight into the robot's motors, batteries, and weapons.

Whyachi!

"The name Whyachi was coined a few years ago by guys in my shop," says Ewert. "It means to bring someone down hard and to inflict massive amounts of pain or damage." Inflict it did. Son of Whyachi went all the way to the top in its very first tournament, beating the crap out of a variety of robots, including Carlo Bertocchini's nearly indestructible Biohazard.

Son of Whyachi's brutal offense also serves as its defense. The robot carries no armor. But to attack it, you must first get through its spinning Cage of Death. The weapon is a five-inch diameter, three-armed spinner, with 10-pound S7 tool steel "meat tenderizers" on the ends. The arms are cross-braced to form a strong cage. The weapon spins up to 700 RPM in two seconds, thanks to dual 12 HP Briggs & Stratton E-Tek electric motors.

Son of Whyachi's walking assemblies are driven by 24-volt Bodine gear motors, over-volted 100 percent to 48 volts. "I have been using Bodines in our industrial equipment for years and they are very durable," says Ewert. The works are powered by Hawker Genesis sealed lead acid batteries.

The robot's frame is made from 3003-H14 aluminum. "It forms and welds nicely," says Ewert. "It is also more 'gummy' when a Kill Saw hits it."

From Sausage Loaders to BattleBots

Ewert is a mechanical engineer, and the owner of Wisconsin-based Westar Manufacturing. Westar designs and builds food processing equipment with such evocative names as "tray denesters" and "sausage loaders."

It took Ewert about three weeks to design Son of Whyachi on the computer using CADMAX software. "One hundred percent of the bots were done in CAD," says Ewert. "It is the only way to achieve a densely packed, strong bot." Son of Whyachi came together in about a month, at a cost of about $7,000 in materials and $24,000 in labor. Sound expensive? Machine shop time is costly. Team Whyachi is run as a business, and the team includes their time and labor—at standard rates—when calculating costs.

BattleBots rules have changed since Son of Whyachi's reign of terror. The weight bonus for walking robots has been reduced from 50 percent to 20 percent, and Son of Whyachi's super-cool shuffling drive system no longer qualifies it as a walker. It is being replaced with more mundane wheels. "We are already fighting at the Super Heavyweight wheeled weight," says Ewert, "so we might as well be wheeled."

Altogether, Team Whyachi boasts a half dozen battle robots, including Whyachi, Whyatica, Y-Pout, and YU812. They also machine and fabricate bot parts for fellow builders "at reasonable prices."

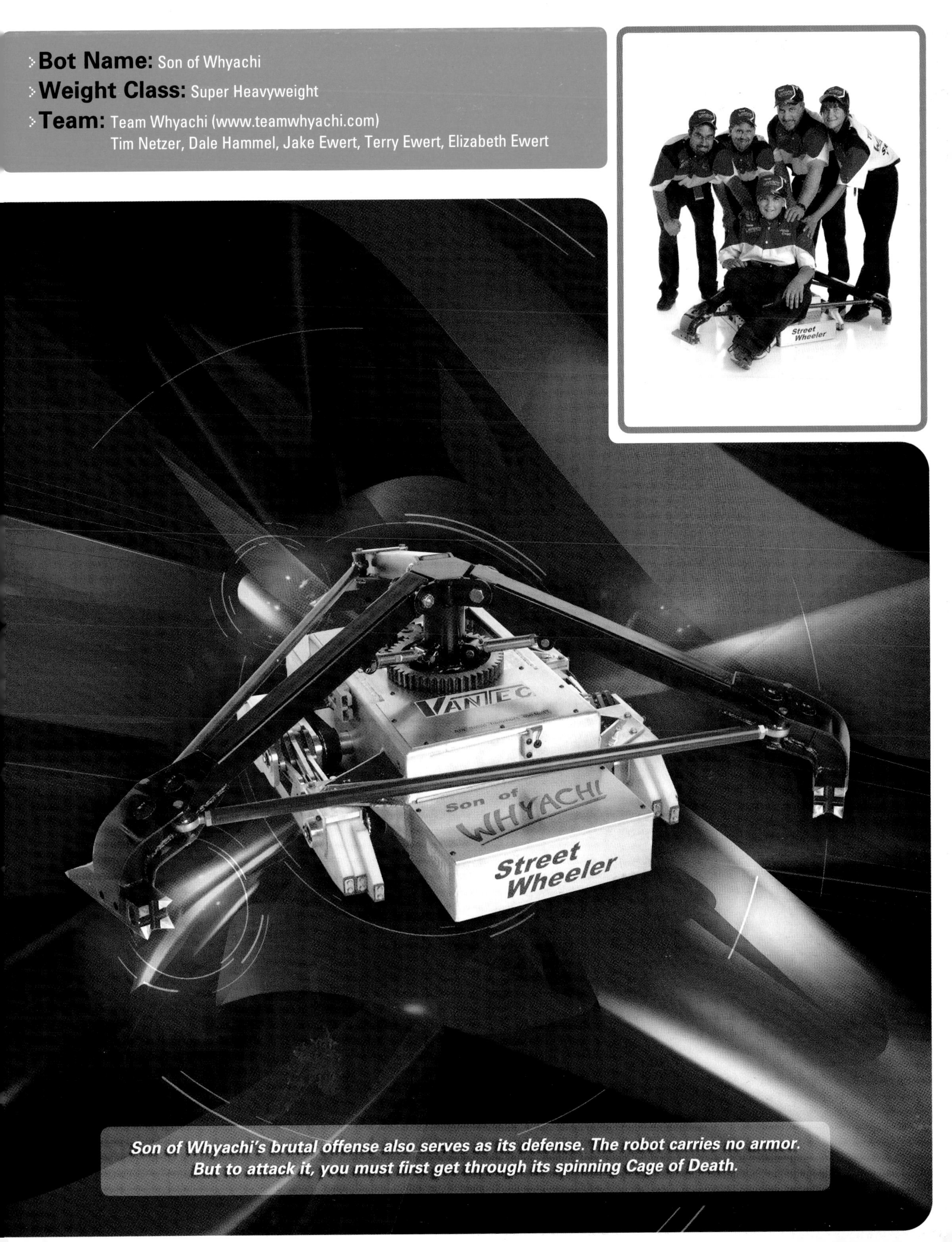

Bot Name: Son of Whyachi

Weight Class: Super Heavyweight

Team: Team Whyachi (www.teamwhyachi.com)
Tim Netzer, Dale Hammel, Jake Ewert, Terry Ewert, Elizabeth Ewert

Son of Whyachi's brutal offense also serves as its defense. The robot carries no armor. But to attack it, you must first get through its spinning Cage of Death.

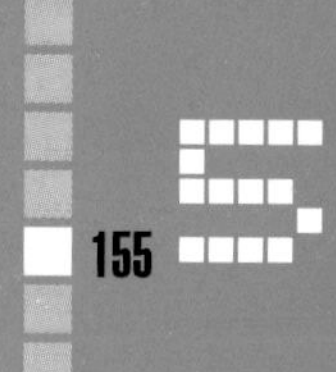

SUNSHINE LOLLIBOT

I'm ready for my close up Mr. Dremel!

Sunshine Lollibot's name is, of course, an homage to the infamously saccharine 1966 Lesley Gore song, "Sunshine Lollipops." It is also, says Lolli's builder Andy Miller, an attempt "to humble all the newbie robots with names like 'Armagedeathbot 3000' who think they're so hot." That's Sunshine Lollibot over there, the pink one wearing the feather boa. Now, don't you feel humbled?

Miller, a mechanical engineering student at the University of Washington, fell into BattleBots like many an idle-handed youth. "I decided I needed a hobby," he says. "I started building the cheap stuff first and, before I discovered how expensive it really is, I was already addicted."

A Big-ass Blade

Miller's initial concept for a BattleBot was simple: "A big-ass saw blade, and a chainsaw motor to run it. Any questions?" The theory behind the concept? "People who'd tried big saw blades before generally used either used metal cutting blades, where the teeth are tiny, or small wood blades, which just stop when they hit something. But add enough kinetic energy to a big blade, and it will actually cut into things."

Miller spent nine months, off and on, designing and building Lolli, including two non-combat iterations. The robot first competed in the May 2001 BattleBots event.

Updates and Revisions

For version 1, Lolli's frame was made of steel. "Steel is cheap, relatively strong, and I could weld it," Miller says. "Version 2 was made of aluminum. It's still fairly cheap, it's stronger than steel for its weight, and my uncle can weld it (thanks!)."

Lolli is armored mostly in Lexan, to help keep her weight down, with some assorted bits of aluminum and titanium placed over especially sensitive areas. Lolli's drive train originally relied on borrowed motors powered by "crappy NiCads from Black and Decker drills," says Miller. "They were fairly lightweight and, above all, cheap." Miller has since updated the bot with EV Warrior electric bicycle motors powered by NiCad BattlePacks.

Bigger, Pinker, Badder

Lolli's big pink weapon, a 24-inch blade from a sawmill, is powered by a 4 HP gasoline engine and spins at around 2000 RPM. Though designed to cut wood, Miller says, "it went through 3/4" thick steel in my first fight." Version 1 of Lolli used a v-belt to drive the weapon, but Miller switched over to motorcycle chain for Version 2. For the next iteration, he's switching back to belts. "Chains have too many disadvantages," he says. "They are harder to keep in alignment and tend to break, rather than slip."

In addition to new belts, Lolli is in for a host of refinements for the next event, says Miller. "Sunshine Lollibot's gonna be back, badder than ever ... and pink paint again!"

What if money were no object? How bad would Lolli be then? "If money were no object," Miller says, "I'd make the frame entirely out of a single carbon 'Bucky Ball' molecule. It'd literally take a nuclear weapon to break the thing."

Bot Name: Sunshine Lollibot

Weight Class: Middleweight

Team: Team Sunshine (www.olympus.net/personal/viviannk/webpics/index.html)
Andy Miller; not pictured, Todd Miller and Viviann Kuehl

"I wanted to humble all the newbie robots with names like 'Armagedeathbot 3000' who think they're so hot."
— Andy Miller

SUPER CHIABOT

The shrub goes full force

Cassidy Wright, now 15, has been competing in robot combat events since 1995. Her first robot was a three-way MultiBot called Triple Redundancy, which she drove with two cousins. Middleweight BattleBot Super ChiaBot is her fifth robot. Her father, Will Wright, is known for his whimsically unconventional approach to robots, but he insists that ChiaBot was all Cassidy's idea.

"At first she wanted to make it with a real shrub she had found in the back yard," says Will Wright, "but I finally managed to convince her that that wasn't practical, and we should use artificial shrubbery."

Building a Better Shrub

ChiaBot competed at the very first BattleBots event in 1999 as a Lightweight. It had no weapons. It was just a shrub. "We determined that wasn't terribly competitive," says Wright, "so we went back to the drawing board and built a Middleweight with a weapon." No longer just a plastic shrub, new and improved Super ChiaBot was a shrub with a spinning disk sticking out of it.

"I helped her in building the robot," says Wright, "but Cassidy drives it."

Super ChiaBot is a MultiBot, a big robot with a little robot tucked inside. At the beginning of the match, the little robot pops out of its docking station, through a trapdoor at the back of the bush. Wright describes Chia's little helper as an "annoyance bot." "It's a little wedge," he says. "It tries to get in front of the other robot and get underneath it, to immobilize it. Then the big robot comes up to attack with the weapons."

Has it worked? "A couple of times," Wright says. "We keep trying it."

Gaze Within

Chia's square frame is made of aluminum angle bar. The bottom is armored in 7075 aluminum. The robot's top is an extremely light aluminum superstructure with the plastic shrubs attached to it. Super ChiaBot has almost no armor, says Wright. "Push apart the shrubs with your hand and you see straight into the motors and the docking bay for the little microbot."

Each of the robot's four wheels is driven directly off of DeWalt 18-volt cordless drill motors powered by gel cell batteries. Both front and rear wheels are toed in, to help ChiaBot turn. The 22-pound weapon disk is actually two flywheels off a Chevy 350, with several inches of after-market S7 steel blades protruding from the edge. It's spun up to about 2200 RPM by an AstroFlight Cobalt 60 electric motor, powered by NiCad batteries. Chia's little helper bot uses Makita right-angle drill motors, powered by a gel cell battery.

Despite Chia's unlikely appearance, it has actually won matches. "If we can keep improving the weapons and adjust the weak spots on the robot, it has the potential to be competitive," says Wright.

After a more competitive shrub, what's next for Cassidy Wright? "She wants to do a sofa," says her dad, "or at least a comfy chair."

Cassidy Wright and Super Chiabot

Bot Name: Super ChiaBot

Weight Class: Middleweight

Team: Robot Action League (www.robotactionleague.com)
Cassidy Wright (pictured below), Will Wright, and support crew

No longer just a plastic shrub, new and improved Super ChiaBot was a shrub with a spinning disk sticking out of it.

TAZBOT

Even among BattleBots, Donald Hutson's Tazbot—with its spinning turret, rakishly tilted wheels, and slinky barbed tail—stands out. "People say it's kind of like a cartoon character," says Hutson. "I like that. It's important to have a lot of character in a robot."

Inspiring Agility

It's also important to win, Hutson admits, "but it always seems to do well. It came down to the agility." In his quest for agility, Hutson borrowed ideas from baseball, the martial arts, and wheelchair basketball.

Tazbot combats Battle Rat

Primary inspiration struck, he says, while swinging a baseball bat around: "I said, 'I want to go up to these guys full speed, swinging.' I needed a turret to give me 360 degrees of offense. I needed wrist action and follow-through, so I made the weapon spring-loaded with a rubber damper. I tried to emulate a person hitting a baseball. I've been wanting to put a baseball bat on Tazbot and go down to the batting cages."

"I also looked at the martial arts," says Hutson. "where a guy with a stick can kick the crap out of a guy with a chainsaw, because he has agility.

"Agility is what Tazbot's whole design is about. If I'm pushed into a wall, I can come around the other way. A robot might get one good hit in, or get underneath me, but they know that weapon's coming around."

Tazbot's weapons are modular; Hutson can add attachments to the rear, the front, and the turret. "I have a plethora of interchangeable tools that people haven't seen—things I'm saving for a special occasion."

A Wider Stance, a Helpful Arm

Tazbot's wheels are cambered at a striking 45 degrees. Why? "Everybody concentrates on forward and reverse acceleration," says Hutson, "but ignores turning. The wider the camber, the easier it is to turn. Plus, it gives me a wider stance and a lower center of gravity. I got the idea from the guys who play wheelchair basketball; they have the same cambered design."

The wheels are direct-driven by two Stature motors. A surplus motor and gearbox spins the turret. The turret arm swings up and down a full 180 degrees, thanks to a Motion Systems linear actuator, rated at 500 pounds. Tazbot can use the arm to right himself, tip other robots, or extract himself from sticky situations. "I've had the front end stuck up on the rail," says Hutson. "I laid the arm on the side of the wheel, lifted the front end off the rail, and drove away."

The robot is powered by high-current SVR sealed lead acid batteries from an unusual source. "They're meant to replace the batteries in cars," says Hutson, "so people who do car stereo competitions can fire up all 20 of their amplifiers at once and the battery will take the load."

Keeping It Pure

Tazbot's come a long way in his six years of competition. Tazbot's frame, originally cheap electrical conduit, is now made of much harder chromoly tubing, with 6061 aluminum armor on the outside and titanium plates underneath. The weapon has evolved into a omnidirectional flipper. And Tazbot has grown nasty, serrated toes. But, insists Hutson, "The things that change are pure. Everything is there for a purpose."

Bot Name: Tazbot
Weight Class: Heavyweight
Team: Mutant Robots (www.mutantrobots.com)
Dawn Berstein, Mike Moore, David Bleakley, David Cook, Donald Hutson
CINCINNATI
NPC
ROBOTICS
"I have a plethora of interchangeable tools that people haven't seen—things I'm saving for a special occasion."
— Donald Hutson

TECHNO DESTRUCTO

Techno Destructo upends Diesector

The first thing you notice about Team Carnage's Super Heavyweight, Techno Destructo, is that she—yes, she—looks a little different from the rest of the pack. "The shape reminds me of a dung beetle," says designer Sean Irvin. While most LaunchBots are basically squarish boxes, Techno Destructo has a distinctive, gracefully curved topside. "I wanted a good combat robot, but I wanted it to be aesthetically pleasing to me," says Irvin. That distinctive curve was originally the idea of Irvin's wife, Amber, and it's more than pretty; it's functional. If Techno Destructo gets inverted during a match, it uses the weapon arm in combination with its curved top to make a nice, gentle, safe roll all the way back onto its feet.

A Very Technical Boy

Irvin holds a master's degree in robotics from Florida Institute of Technology, so he knows from robots. Like many BattleBots designers, Irvin used 3-D computer-aided design software (PTC's Pro/ENGINEER) to design Techno Destructo. "It's great for laying out the robot, going through the motions for all the components, and fit-checking everything ahead of time, so there's no 'uh-oh, this doesn't work.'"

Most BattleBots use PCM radio control systems, like those used in remote control cars and planes, but Irvin chose a controller more suited to his extensive video gaming experience. "I grew up in the age of video games," says Irvin, "so I've always had a joystick in my hand. BattleBot is like a real-life video game." Techno Destructo uses an IFIrobotics controller, with a standard, analog computer joystick on Irvin's end, and a programmable onboard computer on Techno Destructo's end.

Re-Framing

Techno Destructo's initial impressive performance in the BattleBox—six wins in eight matches—didn't stop Irvin from making a few improvements between seasons.

Techno Destructo originally featured a monocoque construction, with no separate frame and all parts bolted directly to the armor. The newest revision sports a separate tubular frame. External armor and internal components all bolt to the frame, making adjustments and repairs much easier. "When the armor gets messed up," says Irvin, "you just build a new piece; you don't have to build a new robot. It keeps everything looking good and ready to fight."

Techno Destructo's original aluminum and steel armor has been replaced with 3/8" Lexan on the top and sides, and 0.090" titanium on the bottom to protect Techno Destructo's innards from the attentions of hungry Kill Saw. "The Kill Saw chew up aluminum and stainless steel like butter," says Irvin. "No mercy at all."

Big Bad Woman

"Everybody likes when bots fly through the air," Irvin says. To that end, Techno Destructo's already impressive 1200 pounds of lifting force has been increased to a scary 10,000 pounds, actuating in 0.0050 seconds, thanks to an upgrade from high-pressure air pneumatics to a 1200 PSI liquid CO_2 system.

Overall, says Irvin, the new Techno Destructo is three times faster, eight times stronger, and a whole lot tougher.

While builders generally regard their creations as male, Irvin insists that Techno Destructo is definitely a woman. "Only a woman could give me the kind of frustration she does. My wife likes to refer to Techno as the other woman in my life. And, at 340 pounds, she's the biggest, meanest, baddest woman I've ever seen."

The robot, not the wife.

> **Bot Name:** Techno Destructo
> **Weight Class:** Super Heavyweight
> **Team:** Team Carnage (www.carnagerobotics.com)
Monte Keen, Richard MacKenzie, Sean Irvin, Amber Irvin, and Don Price

While builders generally regard their creations as male, Irvin insists that Techno Destructo is definitely a woman.

T-MINUS

T-Minus battles Heavy Metal Noise

Inertia Labs' Middleweight LaunchBot, T-Minus, is, if anything, even scarier than its Super Heavyweight sibling, Toro. "Toro can dead-lift the end of a car," says Team Inertia's Alexander Rose. "But with T-Minus, our goal was to increase the strength-to-weight ratio even further. T-Minus weighs only a third of what Toro weighs, but it has half the flipping power. It's a much stronger robot, in relation to its weight."

Less Weight/More Thrilling

T-Minus's current strength-to-weight ratio is, frankly, stunning. T-Minus is a 125-pound robot, with 3,500 pounds of lifting force, thanks to its custom-built high-pressure liquid CO_2-powered pneumatic arm. To handle T-Minus's high pressures, fast actuation, and extreme temperatures (liquid CO_2 systems can get very cold), Inertia Labs custom builds its own oversized, high-pressure valves. On the outside, T-Minus uses an all-aluminum frame and armor—easy to cut, easy to weld. "On the inside of the robot," says Rose, "we put a layer of titanium under anything like wires or electronics, so the saws skip over it if they cut through the armor."

Inertia Labs learned a few lessons from their first LaunchBot, Toro, says Rose, and they put those lessons to work in T-Minus's design. Toro, for example, is not invertible, thanks to its big, humped lifting arm. T-Minus, on the other hand, is invertible; its launching arm folds down flush with its back, allowing it to drive upside down. But T-Minus doesn't stay on its back long. With its powerful launching arm, T-Minus can also self-right, in a *big* way. From flat on its back, T-Minus launches several feet straight up, performs a 540-degree flip-and-a-half, and lands on its wheels. Always a big crowd pleaser.

Inertia Labs was also unsatisfied with Toro's drive train, says Rose: "Toro has a chain drive system, and it's a pain. For T-Minus, we wanted to use a direct-drive gearbox rig." To that end, T-Minus uses National Power Chair (NPC) electric wheelchair motors, each with its own speed controller, hooked directly to its two wheels.

BattlePacks

T-Minus powers those motors with 24-volt NiCad BattlePacks. T-Minus was the first robot to use the increasingly popular BattlePacks. In fact, T-Minus was the prototype implementation for BattlePacks, which Rose designed, along with Steve Hill, who specializes in building NiCad battery packs for R/C cars. "T-Minus was designed around BattlePacks; the first prototype set of batteries was in that robot. They worked so well that we switched all of our robots over to NiCads," says Rose. "We were not getting enough run time out of sealed lead acid (SLA) batteries. NiCads hold their voltage for a longer period of time."

T-Minus has been a very successful BattleBot, and it's definitely going to keep competing, says Rose, "but it's going to have a pretty significant rebuild, including enclosed wheels." T-Minus's wheels are currently exposed. But after watching their Heavyweight BattleBot, Matador, have its wheels ripped off by Marvel of Engineering in Season 4, says Rose, "we decided the future holds all enclosed wheels for us."

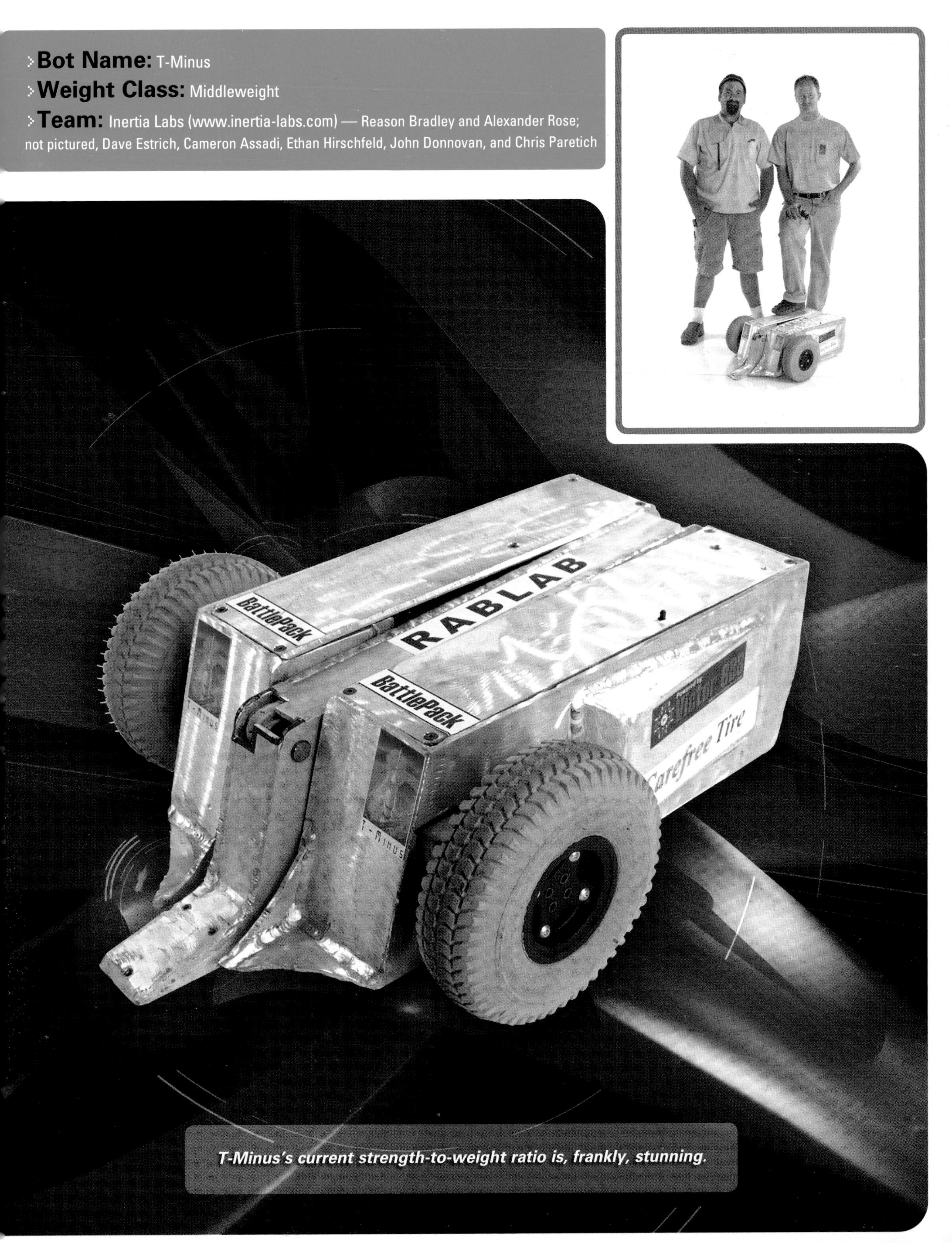

Bot Name: T-Minus

Weight Class: Middleweight

Team: Inertia Labs (www.inertia-labs.com) — Reason Bradley and Alexander Rose; not pictured, Dave Estrich, Cameron Assadi, Ethan Hirschfeld, John Donnovan, and Chris Paretich

T-Minus's current strength-to-weight ratio is, frankly, stunning.

TORO

Toro gets ready

Named for a bull tossing a luckless matador into the air, Inertia Labs' Toro is the reigning Super Heavyweight Season 4.0 champion.

With its huge, trademark lifting arm, arching way up over its back, Toro definitely looks like it means business. But few people at Toro's 2000 Las Vegas debut were prepared for the sight of Toro's opponents sent spinning through the air, courtesy of the 7,000-pound pneumatic rams behind Toro's fearsome flipping arm.

"To my knowledge," says Inertia Labs' Alexander Rose, "we're the first people to really throw robots up in the air."

Son of Rhino

Toro was Inertia Labs' second BattleBot. The first, Rhino, is now retired, but its weapon, a pneumatic ram, lives on in Toro's flipping arm. "We had hyper-developed Rhino's ram mechanism," says Rose. "We could put a hole through a half-inch of aluminum. But then people were putting on three-quarter inches of armor, and we weren't getting through anymore.

"So we thought, if we turned the whole system vertically, we could flip people." And flip they can. Toro's arm actuates in under 1/100th of a second and is strong enough to dead-lift the end of a car.

The high-pressure liquid CO_2 pneumatic system driving Toro's arm is nearly identical to Rhino's, but Toro uses two rams to Rhino's one, for twice the power. "We have to make our own custom, giant, high-pressure valves," says Rose. "There's nothing on the market that'll switch at the pressures and the speeds that we're doing."

Paintball Experience

Where did they get the wherewithal to do that? "I'm an industrial designer," says Rose. "Reason Bradley, my partner, is a machinist. Teammate David Estrich is an electrical engineer.

"Both Reason and I played a lot of competition paintball, and I'd done a lot of design work optimizing high-pressure air and CO_2-based guns. Between us, we were able to come up with our own valving."

To control that custom valving, Toro uses an IFI control system, with an onboard computer. "IFI is a *huge* improvement for us," says Rose. "For Rhino, we had to design our own custom boards for valve sequencing. With the IFI system you can do all that in software."

Toro's exterior is a bit more conventional, made entirely of welded aluminum. "We built Toro in two weeks," says Rose. "We can do that because we are able to cut aluminum on a bandsaw, and weld it together in a day."

Non-Binary Failures

Toro's power comes from 24-volt NiCad BattlePacks. Each of Toro's four wheels is chain driven from a Bosch GPA 750 electric motor, each with its own speed controller. "If one wheel gets jammed up, you still have three functioning wheels," says Rose. Toro's twin rams are also independent, each with its own CO_2 supply.

"We've tried to make all of our failures non-binary," says Rose, "so one small failure doesn't kill the whole robot." A lesson learned in Toro's first competition, where the robot charged through the field—until a battery wire came loose on the radio. "It's a recurring theme for us," says Rose. "At least once per competition one of our robots loses a radio. That's why our latest robot, Matador, has two separate radios, each with its own power supply. That robot could be cut in half and still work."

Bot Name: Toro

Weight Class: Super Heavyweight

Team: Inertia Labs (www.inertia-labs.com) — Reason Bradley and Alexander Rose; not pictured, Dave Estrich, Cameron Assadi, Ethan Hirschfeld, John Donnovan, and Chris Paretich

"We're the first people to really throw robots up in the air."
— Alexander Rose

VOLTRONIC

Voltronic gets under Bacchus

Voltronic's builder, Stephen Felk, has been passionate about robotic combat since he stole into his first Robot Wars way back in 1996. "I went on Friday, and was totally hooked," says Felk. "I came back on Sunday, but it was sold out. So I snuck in. I never do that kind of thing, but I *had* to see that event."

Not long after, he came across an electric wheelchair. "I tore it apart, then took the parts and sort of rearranged them into a fighting robot. It had a mechanism that raised and lowered the chair, so I thought I'd use that for a lifting arm."

A Missed Opportunity

Felk set about building a robot, but even working night and day, he couldn't finish it in time for the next Robot Wars event. "I tried to incorporate a lifter and rotating wheel shields. I had no idea how much time it took to build one of these things.

"Whatever happens to me in this sport it'll never be worse than that: knowing I blew it, knowing I had to wait another year to fight."

In fact, it turned out to be more than two years before Voltronic would finally see combat—if you don't count a little off-the-record nighttime get-together under a freeway overpass.

Remember to Bathe

When the first BattleBots took place in 1999, Voltronic was there and rarin' to go. Felk has made sure it hasn't missed a single BattleBots since: "I typically take six weeks off work to get the robot ready. I can't afford it, but I have credit cards. The credit people love me.

"As I get closer to the event, I weed everything else out of my life. For the last four weeks, it's eat, sleep, work on the robot." During this time, Felk rarely leaves the house, and uses the length of his beard to gauge the time since he last bathed—habits sadly typical of BattleBots builders.

Voltronic is constructed of aircraft aluminum. A base plate and a top plate attach to four longitudinal bulkheads, and all internal components attach to those plates and bulkheads. Voltronic has been using the same Leeson's industrial electric motors from the beginning, powered by good, old-fashioned Hawker SLAs. "Voltronic was designed in 1996," Felk says. "It's way past its prime." Maybe. If so, it's an oldie but a goodie.

A Swarm of Bots

Felk's next goal is to field a robot in all four weight classes. To that end, he's got a new Heavyweight—or part of one—called The Swarm.

The Swarm is actually a MultiBot, a team of three lighter robots that gang up on larger robots. Felk, Jim Smentowski, and Paul Mathus each drive a bot of their own design.

Not all BattleBots fans have warmed to the notion of MultiBots, but Felk is unaffected by the negative crowd reactions. After all, Voltronic is a Wedge, and Wedges are notoriously unpopular with the crowds. "I get booed every time I go into the arena," he says. "I don't pay any attention."

In fact, Mosquito, Felk's robot from The Swarm? It's a Wedge.

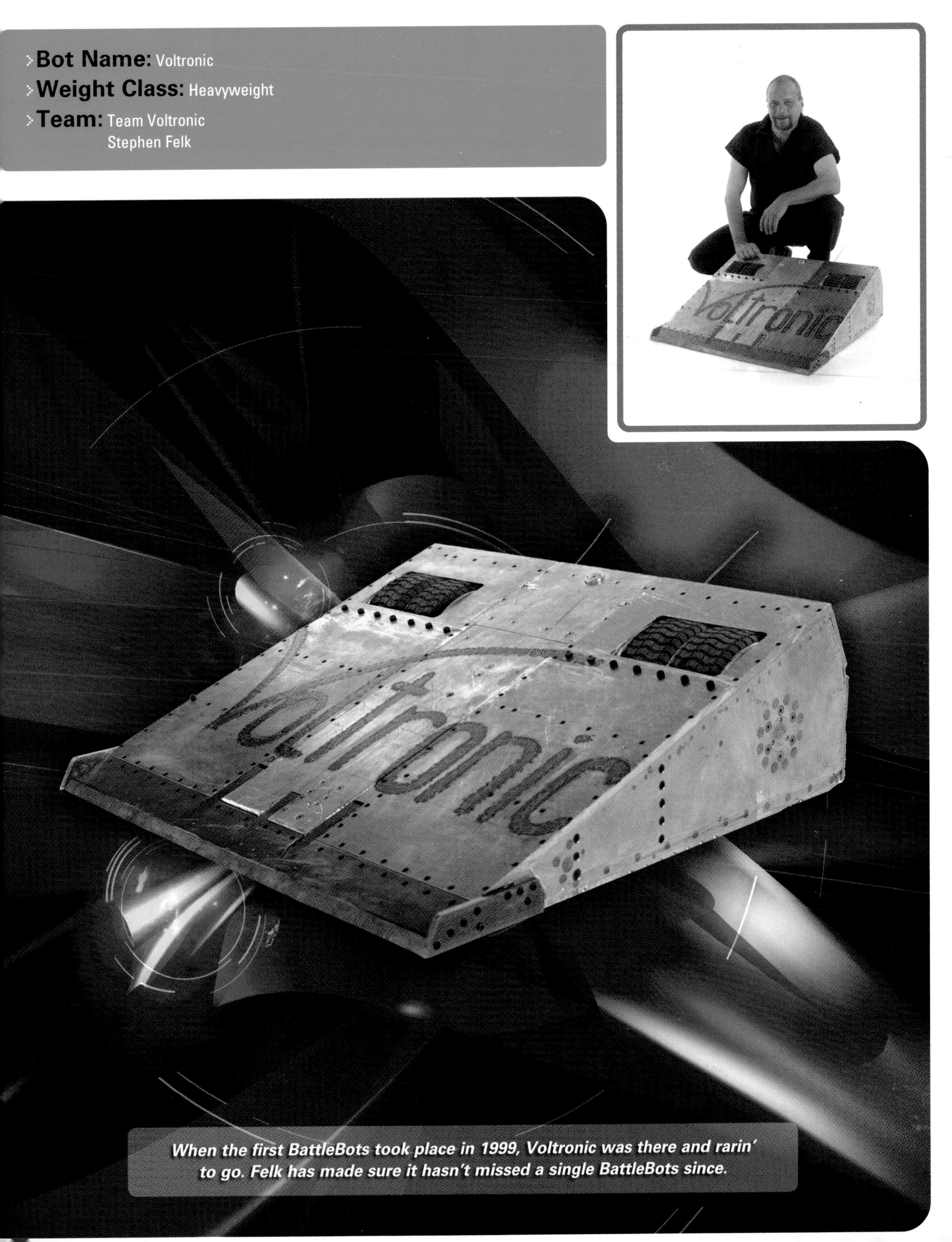

Bot Name: Voltronic

Weight Class: Heavyweight

Team: Team Voltronic
Stephen Felk

When the first BattleBots took place in 1999, Voltronic was there and rarin' to go. Felk has made sure it hasn't missed a single BattleBots since.

WORLD PEACE

Give Peace a chance

Super Heavyweight BattleBot World Peace is, Team Bohica insists, a staunch advocate for the nonviolent resolution of all conflict, although he will reluctantly use force after all other options have been exhausted. His pastel blue aluminum shell is adorned with doves, olive branches drooping from their mouths. World Peace's own beak, however, is a little more threatening, designed to rend and tear into other bots.

World Peace is the "killer app" of the Robotics Division of Bohica Brothers Heavy Industries, primarily brothers Dave and Greg Campbell. Dave Campbell, lead designer of the robot's hydraulic and mechanical systems, specializes in using electric and internal combustion motors in ways the manufactures never imagined possible (or advisable).

How Hard Could It Be?

"I have a couple of theories on tinkering and building things," says Dave. For instance? "If it's already broke, what's the worst that's going to happen trying to fix it?" And? "How hard could it be?" How did he happen to bring these theories to bear on the field of combat robots? "I went to Robot Wars at Fort Mason [in 1997] and figured, 'How hard could it be?' We started designing World Peace at a bar after the show."

There would be no spinning weapons for Team Bohica—too messy. "The philosophy from the beginning," says Dave, "has been a weapon that could fatally disable another robot, but with *finesse.*" The result, World Peace's beak, is an outsized set of tin snips, driven by 5000 PSI hydraulics.

Custom Radios and Gasoline Engines

Dave's brother and teammate, Greg, handbuilt the team's radio control system, using his own "that didn't work, what happened?" system of engineering. Greg's circuit boards won the PC Express design contest in *Nuts & Volts* magazine. The resulting one-of-a-kind control box looks like a bastardized WWII military radio with joysticks.

World Peace's bottom cover is titanium. The top covers are 6061 aluminum. The frame and chassis are rectangular tube steel. All pretty standard stuff. But under the covers, World Peace is a little different, one of the few robots using a gas engine for propulsion. "We keep wondering if everyone else knows something we don't," says Dave. World Peace's drive train is powered by a Stihl 088 8.5 HP chainsaw engine, turning a Vickers vane pump. Mr. Peace's four wheels are chain-driven off of Charlyn hydraulic motors.

So far, World Peace has had some problems staying on its wheels and putting the bite on the bad guys, but not to worry, says Dave. "What we have now is just the prototype. The design is starting to take shape. Everything will be changed in the next version."

It's All About the Good Times

Win or lose, Team Bohica never forgets that BattleBots is all about having a good time. "At the second Vegas competition, Stan, our tequila guy, was passing out shots to other teams in the parking lot by our trailer. We were approached by another competitor who made the comment, 'If you guys didn't drink so much, you would do better.'

"Sure. But at what cost?"

Bot Name: World Peace

Weight Class: Super Heavyweight

Team: Team Bohica (www.bohicabros.com) — Stan Kitaoka, Greg Campbell, Ed Bitondo, and Dave Campbell; not pictured, Ed "Pinky" Spencer and Ron Bitondo

"The philosophy from the beginning has been a weapon that could fatally disable another robot, but with finesse."
— Dave Campbell

ZIGGO

You've just been Ziggoed

There was a time when Lightweights were generally considered incapable of inflicting real damage on each other. That time was before Ziggo. Ziggo destroys opponents, sending chunks bouncing off the walls and even the ceiling, 26 feet above. With an impressive 18 wins in 21 bouts—16 of them by knockout—it's no wonder Ziggo is the number one robot in its weight class.

Ziggo's builder, Jonathan Ridder, named the devastating little bot after his misanthropic cat, Ziggy (now deceased), and Blendo, the destructive robot that inspired the design. Ziggo's elegant, effective construction belies Ridder's lack of formal training in engineering or robotics. Formal training or not, Ridder is a robot combat veteran, building and competing since 1995.

Maximum Carnage

His goal for Ziggo's design? Maximum carnage. "I wanted to have a truly destructive robot," says Ridder. "The only way to do a lot of damage is to transfer a large amount of energy from your robot to the other robot. I decided that the best way to store up a lot of energy was in the form of a large flywheel."

That flywheel became Ziggo's armored spinning top. Made from up to six layers of 1/16" steel, Ziggo's top is both its defensive armor and its offensive weapon. Two Aveox brushless electric motors provide about 1.5 HP each to spin the 25-pound lid up to its top speed of 2000 RPM. Ziggo is also a highly mobile robot, able to evade opponents for the few seconds it takes its top to come up to speed.

Ziggo is built on a base plate made of hardened #01 tool steel, beneath a second base plate of aluminum—actually Ziggo's original base plate. Aluminum ribs are screwed to the base plates to stiffen and strengthen the chassis. The spinning top mounts to a spinning control shaft, also screwed to the chassis.

Ziggo uses NiCad batteries to provide the current its weapon and drives need. "NiCads are the best at providing large amounts of current in a short amount of time," says Ridder. Ziggo uses DeWalt 18-volt cordless drill motors and matching gearboxes, run at 24 volts, to drive two of Ziggo's four wheels. The other two wheels are driven off the first via timing belts.

Better and Better

"Ziggo is upgraded before each new competition," says Ridder. Over the years, the top has gone from 8 pounds to 25 pounds, and the single 1 HP motor that originally spun it has been replaced with dual 1.5 HP motors. "The weapon's gearbox has been redesigned three times to make it more reliable," says Ridder, "and I will be adding a forced air cooling system for the motors and controllers that spin Ziggo's top.

"Each change has been made in order to make Ziggo more powerful or more reliable. Ziggo is pretty well optimized at this time."

Absolutely. Just ask anyone he's destroyed.

- **Bot Name:** Ziggo
- **Weight Class:** Lightweight
- **Team:** Team Ziggy (www.teamziggy.com)
  Jonathan Ridder and Douglas Ridder

"I wanted to have a truly destructive robot."
— Jonathan Ridder

PART THREE

HOW IT

6 — HOW TO BUILD A BATTLEBOT

Come on. Admit it. The first time you saw BattleBots on TV you said, "I should build one of those." Maybe you were stirred by Nightmare's ability to destroy. Or you might have been impressed by Biohazard's elegant design and construction. Perhaps you were a little turned on by Mauler's pure beastliness, or inspired by the way rookie robot New Cruelty made it all the way to the finals without any weapons at all. So now you're thinking, hey, this robot fighting stuff is for me.

Never built a fighting robot before? That's okay; few people have. In fact—please don't tell anyone—I've never built a fighting robot, either. That's why I turned to experts in the field, securing advice and words of wisdom for us tender, first-time bot builders.

'S DONE

Building Your First BattleBot, with Christian Carlberg

So you want to build a BattleBot of your very own, but you don't know where to start. Well here's the skinny from veteran BattleBot builder Christian Carlberg. Come along as Christian guides us through the process of building a BattleBot. It's not as hard as you think, and it's a lot of fun.

The very first step is thinking up a cool name for your robot. "It's very important to have a cool name," says Christian. "How else are you going to strike fear and terror into your opponent?"

How Big Is My BattleBot?

The next step is deciding how big your BattleBot will be. As a general rule, the bigger the robot, the more it costs. But Lightweights present their own problems; it's hard to squeeze a combat robot down to 60 pounds.

A Middleweight robot is a good first-time project. If you find lots of cool, lightweight motors and materials, go with it and build a Lightweight. If your robot's turning out too heavy, all right, make it a Heavyweight. There's really not a whole lot of difference between basic robots of different weight classes. Heavier robots use bigger motors, bigger batteries, and more steel and less aluminum.

If you're a first-time builder, stay away from hydraulics, pneumatics, and other super-cool, super-complicated technologies. In fact, for your first time, you should concentrate on building a strong, reliable, two-wheeled base. Worry about building your super-duper all-destructive weapon later, but if you do want to have a weapon, now or later, leave plenty of room to bolt the thing on.

Understanding Your Inner Drives

It's important that you understand just how a BattleBot moves. Most BattleBots employ scrub, or tank-style, steering. Two-wheeled OverKill, four-wheeled Diesector, and six-wheeled Minion all use scrub steering. If you want your BattleBot to move forward, the wheels on the left and right sides spin forward. If you want the robot to back up, the wheels all spin backward. If you want to turn your robot left, the right wheels spin forward and the left wheels spin backward.

Some robots use "car steering," but this is a more complicated method to build and for this tutorial we will concentrate on scrub steering.

Finding Your Motors

To drive your robot's wheels (or treads), you'll need permanent magnet DC motors, rated for somewhere in the neighborhood of 24 volts (24 volts is ideal). AC motors, like those in your office fan, will not work. Three good choices for the first-time builder are cordless drill motors, rebuilt wheelchair motors, and surplus motors.

Surplus Motors

Your cheapest source of robot motors is your local surplus store. Don't have a local surplus store? You can find an online store on the Web. One of the best out there is C&H Sales Surplus Company (www.aaaim.com/CandH/).

You *may* be able to find out the specs of surplus motors—voltage, amp draw, etc.—but you often can't. So how do you know if a given motor will work in your robot? There are a few things to look for.

One tip-off that you've found an appropriate permanent magnet motor is that it's, well, magnetic. Hold a screwdriver or piece of ferrous metal up to the motor can, and you should be able to feel the attraction of the motor magnets, even when the motor's not running.

The very first step is thinking up a cool name for your robot.

BATTLEBOTS®

got steel?

more on how to build

Photo courtesy of David P. Gilson, Team Critter

A budding bot builder

The motor should have two leads, or wires. If a motor has three or four leads, don't bother with it—it's a series-wound motor. They're a poor choice for a first-time builder.

Heft the motor. As a rule of thumb, the more a motor weighs, the more powerful it is. Yes, for enough money you can buy very light, powerful motors, but a surplus motor that weighs two pounds isn't likely to have much oomph. Look for motors in the five- to ten-pound range.

Figure that your motors will account for around 15 percent of your finished robot's total weight. If you're building a Lightweight robot with two motors, those motors should weigh four or five pounds apiece. If the motor weighs ten pounds and you want to use four of them, you'll have to build at least a 120-pound Middleweight.

Gear Heads

A motor all by itself is no good to you. You can't attach a wheel directly to an electric motor's output shaft; the shaft is too weak and turns too quickly.

24V electric motor.
This motor does not include a gear head.

What you want is a motor with a gear head (or gearhead). The gear head contains a set of gears, specifically designed to work with that motor, which reduce the speed and increase the torque of the output. If you buy a motor without a gear head, you'll have to try to find a gear head to fit it or build your own reduction gearing setup—definitely not recommended for the first-time builder.

How do you identify a motor with a gear head? It's pretty easy. An electric motor is basically a cylindrical "can," with the output shaft extending from the exact center of one face of the cylinder. In a gear head motor, the output shaft usually does *not* extend from the center of the cylinder; it is offset, canted, or even set at right angles to the cylinder.

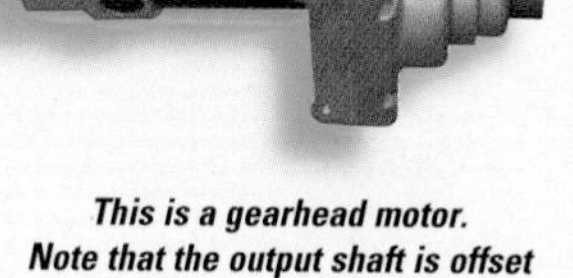
This is a gearhead motor. Note that the output shaft is offset from the center of the can.

If you're still not sure, most surplus places will allow you to hook up your motor to a voltage supply and try it out. Attach the motor to 24 volts and watch how it spins. A motor that spins too fast to see probably doesn't have a gear head and, in any case, isn't appropriate. You're looking for an output shaft that rotates at around three to six revolutions per second (180 to 360 RPM).

Another quick measure of a motor is the size of its output shaft. The shaft is matched to the motor's output power. The bigger (thicker) the shaft, the beefier the motor. A usable gear head motor will have at least a 1/2" diameter output shaft. A motor with anything less than a 1/2" diameter output shaft, gear head or not, isn't strong enough to use in your BattleBot. An output shaft of 3/4" or 1" indicates a very "torquey" motor, indeed.

Look for gear head motors with mounting holes, so you can bolt them directly to your robot.

If you find some surplus motors that you like, buy 100 percent spares. In other words, if you're building a two-wheel robot, buy four motors; if your robot needs four motors, buy eight. Surplus motors are usually cheap, and you may not be able to find replacements when you need them later on.

Drill Motors

Many builders use cordless drill motors, from DeWalt, Makita, or other manufacturers, to drive their robot. These motors are rated anywhere from 9.6 to 24 volts, they come with gear heads, and you can pick up everything you need,

> *"What is the most important single factor [in winning]? It is something you rarely hear discussed. An idea so simple that it would be very unlikely to make most people's top-ten list of secrets to success. Yet it is a secret that most of the veterans know. Are you ready? Here it is: Finish your robot before you come to the competition!"*
>
> *— Carlo Bertocchini*

including batteries and chargers, at your local hardware store. Buy the 24-volt drill motors, even though they are more expensive than the 9.6-volt variety.

One disadvantage of these motors is that they do not have mounting holes for easy installation in your robot. If you go with cordless drill motors, you'll have to make your own motor mounts or buy some from another builder.

Wheelchairs

Christian's recommendation for the first-time builder is to use wheelchair motors. Find yourself a junked wheelchair, or a source of rebuilt wheelchair motors, such as National Power Chair (www.npcinc.com). Wheelchair motors typically run at 24 volts, have a gear head already attached, and have mounting holes. This is a perfect combination for building a robot.

Wheelchair motors are also strong. They are meant to move 300-pound people, so they can easily move your 120-pound robot around.

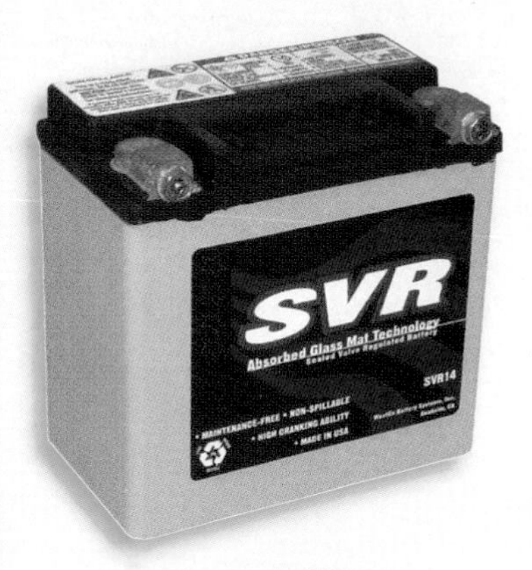

14 amp-hour, 12-volt SVR battery with screw-down terminals

Batteries

Any decent 12-volt battery is fine to start with. You can always upgrade your battery. Head down to the local RadioShack or Wal-Mart store and pick up some 12-volt batteries (at least two). Look for sealed lead batteries, the kind used in wheelchairs and scooters, rather than "wet cell" batteries like the one in your car. Wet cell batteries are all right for bench testing, but they are dangerous once you start driving your robot around, and strictly *verboten* in BattleBots.

How much battery power you need is determined by the size of your robot and its motors. Batteries are rated in amp-hours. A 10 amp-hour battery would supply 10 amps for an hour, or one amp for 10 hours—in theory, at least.

As a rule of thumb, your Middleweight robot will need around 8 amp-hours of juice. Lightweights need 5 to 6 amp-hours, Heavyweights around 13. Keep in mind that this is only a general rule of thumb. If you have high-performance motors, you will need better batteries.

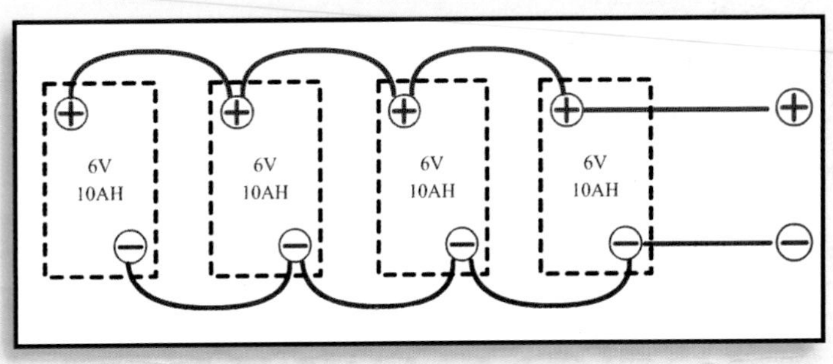

Four 6-volt, 10 amp-hour batteries wired in parallel = 40 amp-hours at 6 volts

Buy batteries with screw-down terminals, if you can get them. If you get batteries with slide-on connectors, make sure you tape the battery leads down to the battery, so the connectors can't slip back off.

You need two 12-volt batteries, wired in series, to produce the 24 volts your robot needs. Batteries are wired in series to double their voltage. Two 12-volt, 5 amp-hour batteries, wired in series, produce 5 amp-hours at 24 volts.

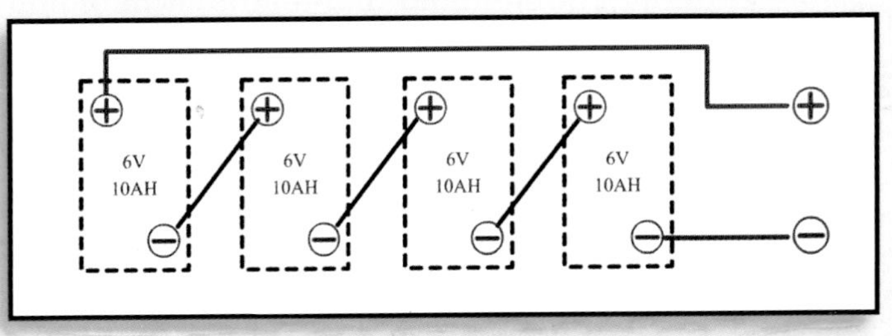

Four 6-volt, 10 amp-hour batteries wired in series = 10 amp-hours at 24 volts

Wire batteries in parallel to double their capacity. Two 12-volt, 5 amp-hour batteries, wired in parallel, produce 10 amp-hours at 12 volts.

You can mix and match series and parallel wiring to meet your needs. If all you find are 6-volt, 6 amp-hour batteries, for example, wire two sets of two of the batteries together in series, to produce 6 amp-hours at 12 volts. Then wire the two pairs together in parallel for 12 amp-hours at 12 volts.

Photo courtesy of David P. Gilson, Team Critter

A freshy welded frame basks in the sun

It's good to have batteries that'll last in your robot for six to seven minutes under load. Five minutes is cutting things a little fine—Rumbles last five minutes. But if your robot runs ten or twelve minutes on a charge, you might consider switching to smaller batteries and using the saved weight elsewhere (such as in more armor). When in doubt, err on the side of caution. Better ten minutes of power than two minutes. The best way to know how long your batteries will last is to drive the robot around with your brother-in-law sitting on it, or push it up against a curb to stall the motors, and see how long your batteries last.

Count on about 15 percent of the robot's weight being in batteries. Batteries provide excellent ballast; you can move them around to adjust your robot's balance and center of gravity.

Photo courtesy of Rick Nickel, Team Redneck Robots

You can build your own battery packs

Over-volting

Many, if not most, BattleBot builders over-volt their motors (run them at higher than their specified voltage). This is a cheap and easy way to get more power out of your drive train. Over-volting by 50 percent (running 24-volt motors at 36 volts) is usually safe, especially if you're using new, high-quality motors that are rated for continuous duty.

If you're buying hot-rod motors, run them at their recommended voltage. If you're using surplus motors... who knows? Try it and see if it works.

Some people even double the voltage to their drive train, running 24-volt motors at 48 volts, but builder beware! You can smoke your motors.

Photo courtesy of Jim Sellers, Robot Action League

The insides of Afterthought

How Many Amps?

Whether figuring out how much battery power you need, or selecting appropriate gauges of wire, it's nice to know just how much current your motors draw. Under load, your motors will probably pull 10 to 20 continuous amps. If you want to know for certain, measure the amp draw on your motors with a multimeter. A multimeter is also good for measuring voltages and checking wire connections for breaks. It's a useful tool to have around your shop.

Note that the more load an electric motor is under, the more power it draws. To really test your motor's current draw, you need to put a load on it. When you're bench testing, you can push a board against the wheel, for example. When your frame and drive train are finished, you can put the motors under load by setting the robot on the ground and running it around or pushing it against a wall for maximum strain.

If you don't have a multimeter, or don't want to chase around after a moving robot with one, get yourself a variety

The best way to know how long your batteries will last is to drive the robot around with your brother-in-law sitting on it, or push it up against a curb to stall the motors, and see how long your batteries last.

of fuses—5 amp, 7 amp, 10 amp, 12 amp, 20 amp—and hook them, one at a time, between the motor and the battery. If the fuse blows, you know you're drawing more current than that. Swap in bigger fuses until they stop blowing. If you blew the 7 amp fuse, but not the 10, then you know your motor is drawing more than 7 but less than 10 amps.

The Speed Controller

You've got your motors and batteries. You've hooked up a motor to the battery and the motor turns. You also know about how many amps you are drawing under load. Now it's time to install the electronics that will control the finished robot's movement: the speed controller.

Electric current coming from your batteries to your motors runs through the speed controller. The speed controller varies the amount of voltage it lets through, controlling the motor's speed. The higher the voltage, the faster the motor turns. Speed controllers can also change the *polarity* of the voltage feeding the motors. By reversing the polarity, it reverses the direction of the motor's rotation. The voltage and direction of the current feeding your motors at any time are determined by a signal coming from your radio receiver in response to your controls.

The vast majority of BattleBots use speed controllers from either Vantec or IFI Robotics. Make sure your speed controller will handle the voltages and amperage you plan to draw. Almost any speed controller can handle 24 volts, but double-check if you're thinking of running at 36 volts, 48 volts, or more.

Photo courtesy of Rob Sica, Team Diginati

Pile of parts: motors, batteries, and speed controller

Some controllers will allow you to *independently* control the left side motor and the right side motor at the same time. You need only one of these controllers to control both sides of your robot independently.

Most controllers, on the other hand, will control only one motor at a time. You would need two of these controllers, one running the left side and the other running the right side.

Sometimes you can double-up motors on one controller. Perhaps you're building a four-wheeled robot, with each wheel driven by its own motor. Instead of buying four individual speed controllers (one for each motor), you can buy one for both the left motors and one for both the right motors. Tie the wires from both left motors together and attach them to one speed controller. (This method is not recommended if your motors draw a lot of amps.)

Some speed controllers provide onboard mixing. Instead of inputs for right and left motors, these speed controllers have inputs for throttle and steering. We'll talk more about mixing later, but this is a big plus if you are planning to use a pistol-grip–type radio.

Expect to spend $300 to $500 on your speed controller(s).

Your Radio

It's time to buy your radio. You'll need a radio that's rated for ground use. Any radio designed for RC cars and trucks will work. These radios are usually pistol-grip style, with a trigger for the throttle and a steering wheel mounted on the side.

Futaba 8UAP 8-channel dual joystick radio

Many robot builders prefer aircraft radios. Aircraft radios are generally higher quality than car radios and have more channels and more features. While a pistol-grip radio typically has only three channels, an aircraft radio will have six or more.

If you do choose to buy an aircraft radio, you'll have to have it converted for ground use, either by the manufacturer or at your local hobby store.

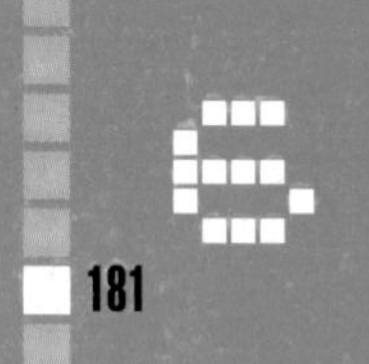

IFI robotics makes a 900 MHz digital radio controller that's becoming increasingly popular among BattleBot builders. These digital radios avoid the frequency conflict issues of traditional RC radios, and the IFI receiver is highly programmable, allowing you to put a lot of intelligence onboard your robot. IFI systems are an expensive choice, especially for a first-time builder, but they are very flexible.

Expect to spend between $150 and $600 dollars on a radio. Really high-end radios can run you $1,200 or more.

Mixing

If you're using a pistol-grip radio—or any transmitter with one control for steering and one control for forward/reverse—you'll want to make sure that either the radio itself or the speed controller you choose provides onboard mixing.

Photo courtesy of Dave Campbell, Team Bohica

Custom-built radio from Team Bohica

Without mixing, each channel of your radio controls one motor. If you have a transmitter with two joysticks, one for each side, that's fine. Both sticks forward would send full power to both motors, moving the robot forward at top speed. Both sticks backward would send the robot backward. One stick forward and the other back causes the robot to spin in place as its motors turn in opposite directions. This is called tank (or scrub) steering.

But if you have a pistol-grip controller, one channel is intended to control the forward and backward motion, not of a single motor, but of the entire robot. Likewise, the second channel controls the side-to-side steering of the entire robot. When you manipulate the steering control, the speed controller must vary voltages to *both* motors, by different amounts, to turn the robot.

Mixing translates the forward/backward and turn-left/turn-right signals into individual voltages for each motor. If you are going full speed ahead and turn right, for example, the speed controller will lower the voltage going to the right motor(s), causing the robot to veer in that direction.

Use a pistol-grip transmitter without any mixing and you'll wind up with the trigger controlling one motor and the steering wheel controlling the other. Not a good situation.

With mixing, you can also implement single-stick mixing for aircraft style radios: push the stick forward, the robot moves forward; push it right, the robot turns clockwise; and so forth.

You can get mixing built into your speed controller or built into high-end radios. You can also buy separate mixers from hobby stores that plug into the speed controller.

Radio Failsafe

To comply with BattleBot regulations, you'll need a radio with a failsafe: if you lose the signal at any point, the receiver sends a zero signal to the speed controller and the robot just stops. This is a feature of some, but not all, radios. Make sure the radio you buy has it.

Implementing the failsafe is a matter of properly programming the radio. Check your manual for details. Don't show up at a tournament without the failsafe working; the safety inspectors won't let you compete.

Photo courtesy of Rick Nickel, Team Redneck Robots

The sad, sorry state of a bot builder's living room

Don't show up at a tournament without the failsafe working; the safety inspectors won't let you compete.

Testing the Radio

When you take your new radio home and open the box, you should find your transmitter (the thing you actually hold), a receiver (a little box with an antenna wire coming off of it), the receiver battery, and a charger.

Wire up everything nice and neat

Photo courtesy of Rob Sica, Team Diginati

The radio may also come with a few servos, small boxes with little wheels on them. You typically don't need the servos for your BattleBot, but if you've got them, you can use them to make sure your radio works.

The radio receiver box will have connectors for the servos to plug into, one for each channel on the radio. Plug a servo into any of those connectors, plug in your receiver battery, and turn on your transmitter. One of the controls on your transmitter should now cause the servo to move. If you have a nine-channel radio, you might have to do a little searching, but keep trying different controls on the transmitter and plugging the servo into different connectors, until you get it to respond. (If you get really desperate, you can always read the directions that came with the radio. They'll tell you which connector is which.)

Once you get the servo to respond to your control, you know the radio's working. If you pull the servo out of that connector on the receiver, and plug the robot's speed controller into that same connector, the same stick movement (or whatever) on your transmitter should now cause your motor to move. Let's hook up the speed controller and see if it works.

Plugging It All Together

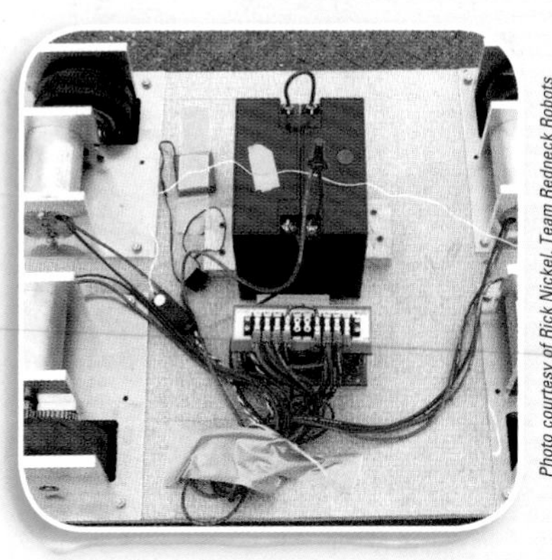

Test fitting the parts on a wood frame

Photo courtesy of Rick Nickel, Team Redneck Robots

Plug your batteries, motors, and radio receiver into the speed controller. The speed controller should be clearly labeled, but some controllers have multiple connectors for motors and batteries. Consult the instructions if you're unsure about what goes where. Both the Vantec and IFI systems plug directly into your receiver box, although the IFI will require a servo extension cable (provided). Don't worry too much about the motor wires themselves. If you put one or both on backward, you can swap them around later, no harm done. However, make sure that you properly connect the battery positive and negative to the speed controller or you might have to buy another speed controller.

When everything's plugged in, turn on your transmitter. Move those controls around and watch your motors spin in response. Congratulations! At this point, you've got the guts of a BattleBot. It's time to lay out the frame.

Motors and wheels mounted at an angle

Photo courtesy of David P. Gilson, Team Critter

Laying Out the Frame

You've got all the parts that will go inside your robot. Now to figure out how best to fit them together, and build a robot around them.

If you are familiar with CAD software, by all means, design your robot on the computer if that's your inclination. But CAD isn't needed. Just take all your parts and put them on the floor or, better yet, on a large worktable or workbench. Move things around until you've got a configuration that you like.

Now outline that configuration in chalk, or tape, or wood blocks. Measure the results; that's how much room you've got to work with.

Attaching the Motors

The type of frame you build and how you put it together are dictated in large part by your motors. You have to figure out how you're going to attach them to your robot. How can you bolt your motors to one piece of metal, then attach that piece of metal to other pieces of metal, and so on, to form some kind of box?

Look at the mounting holes on your motors. Are the mounting holes parallel to the output shaft? That is, does the motor's output shaft go through the same plane that the motors bolt to? If so, you can mount the motors on the side of the robot. You might fold some metal—say, some 1/8" thick 5052 aluminum—into a box, then weld it up, punch some holes in it, mount the motors to the inside of the wall, and attach the wheels.

If the mounting holes are at a right angle to the output shaft, you can mount the motors to a solid metal bottom plate (or top plate) with the output shafts overhanging the plate. This is probably the simplest solution for the first-time builder. You can bolt your other components—batteries, electronics, weapons—to the same baseplate. You don't need to build a frame at all.

Aluminum, Steel, or Wood?

What kind of material will you make your robot out of? Two popular choices are aluminum and steel. A lot of great bots are built with each (or both). The two metals are almost the same in terms of strength for weight—you'll either use thinner steel or thicker aluminum. Either is available in a variety of different shapes, channels, L's, and tubes. Steel is easier to weld. Thicker aluminum will buy you time as the BattleBox saws try to cut through your robot.

Photo courtesy of Rob Sica, Team Diginati

Frame elements in the jig and ready for welding

Don't have any steel or aluminum handy? Run down to the hardware store and buy galvanized angle iron with prepunched holes and slots, and build a frame out of that.

You can even use wood. Wood may sound silly, but wood is not as intimidating to use as steel or aluminum. It's easy to cut, easy to run screws into, and very cheap to use while you're figuring out all this robot-building stuff.

Photo courtesy of Seam Irvin, Team Carnage

A freshly painted green frame for Techno Destructo

What kind of material will you make your robot out of? Two popular choices are aluminum and steel.

Prototype

If nothing else, you can use wood components in your prototype, and then upgrade to metal later. Remember: your first robot is always a prototype, whether you intended it to be or not. You're always going to say, "Gee, I wish I'd done this part a little differently." When in doubt, try it. The sooner you try things and make mistakes, the sooner you'll learn what works.

The only thing that's worth spending good money on is your radio and speed controller. Those will be with you a long time, through many robots.

Photo courtesy of David P. Gilson, Team Critter

CAD – Cardboard Aided Design

Photo courtesy of Rob Sica, Team Diginati

Always wear your safety glasses!

Choosing a Battery

There are a lot of battery choices, but, in reality, almost all BattleBots are powered by one of three types of batteries: sealed lead acid (SLA), nickel cadmium (NiCad), or nickel metal hydride (NiMH). Which battery is right for you depends on your particular application and on your budget. I turned to Steve Hill of Robotic Power Solutions (www.battlepack.com), manufacturers of NiMH and NiCad BattlePacks, for some tips in choosing among them.

▲ Sealed Lead Acid

"Sealed lead acid is a very good way to start," says Hill. "Lead acid batteries cost a lot less, and they use simple charging technology. You can go to Wal-Mart and buy a 10-amp battery charger to charge them with for $30." NiCad and NiMH batteries, on the other hand, require expensive constant current charger setups that'll set you back $150–200. The downside is that sealed lead acid batteries are big, bulky, and heavy, and it takes two batteries to make 24 volts.

Sealed lead acid batteries can put out lots of current, but they have a steep discharge curve. As they drain, the voltage they supply is dropping all the time, especially if you're drawing lots of current. "When you put a high current on a lead acid battery," says Hill, "you reduce its capacity by 45 to 65 percent. If you have a 10 amp-hour battery, you're down to 5 amp-hours."

▲ NiCad

NiCads, on the other hand, have a long, flat discharge curve. They'll run your motors at the same speed until right at the end, and they'll give you 90 percent of your current, even at very fast discharge rates.

NiCads are also much lighter than sealed lead acid batteries, weighing in at perhaps 50 percent of the heavier batteries.

▲ NiMH

NiMH batteries are even lighter than NiCads, offering about a third more capacity for weight. But NiMH batteries have a much higher internal resistance than NiCads. As a result, they are only about half as good as NiCads at dumping large amounts of current quickly.

▲ It Depends on Your Application

"The right battery depends on your robot and your strategy," says Hill. "If you have a Lightweight or a Middleweight, and you're driving it around with minimal stall situations, NiMH is a good solution. It's a lighter, smaller cell with a good capability.

"If you've got a spinning weapon that's going to stall a lot, you want to use NiCads. We say if you're going to pull 40 continuous amps, you need to go NiCad."

Internal Framework: The Box Within the Box

In addition to the motors, your internal frame must hold your batteries, radio receiver, speed controllers, and other components.

Shock-mount your batteries and make sure that whatever they're shock-mounted to is secure, that it can take some major forces and not sheer itself loose. To play it safe, you should put the batteries and the electronics in their own enclosed boxes, and then shock-mount those boxes to your frame with rubber. Your local RadioShack carries premade metal boxes in different sizes.

Photo courtesy of Rob Sica, Team Diginati

Fitting everything within the frame

Wire everything up nice and neat; you'll be working in here a lot. Bring all the wiring together at convenient, central locations: terminal blocks. You can buy a sturdy terminal block at RadioShack or an electronics hobby store. Bring the positive leads from your batteries and speed controllers to one terminal block, and the negative leads to another.

BattleBots rules and general safety require you to have an easily accessible on/off switch on the outside of the robot. This switch must cut power from your batteries to all motors and weapons, shutting the robot down completely in the event that it tries to "run away."

Also remember that, unless you're using transparent polycarbonate armor (such as Lexan), your receiver antenna will have to be mounted so that it extends outside the robot's frame and armor.

Wheels

BattleBots use wheels from a wide variety of sources: wheels from go-carts, wheels from industrial equipment, wheels from lawn tractors, mystery wheels from surplus houses.

This pneumatic-style tire is filled with foam for better durability

Your local hardware store carries wheels for wheelbarrows, lawn tractors, and so forth. Drive down and see what they've got in stock. Look for a wheel that is the right diameter for the robot you had in mind, and that has a four-bolt hole pattern around the center. Wheels with solid rubber or foam-filled tires are best, but you might not be able to buy these at your local hardware store. Pneumatic (air-filled) tires aren't very good in the arena; you can have them filled with foam by such companies as American Airless (www.americanairless.com).

Most hardware store wheels are free-wheeling; they have a hub and bearings that are designed to spin freely about the axle. You can take off this hub and throw it away; a free-spinning hub won't work on a powered wheel.

A nylon wheel with solid rubber tire

Whatever wheel you choose, your task is to lock it mechanically onto your motor's output shaft. You can weld your wheel to the shaft, but that's not a good idea. A better idea is to build your own hub, probably out of aluminum, that matches the hole pattern in your wheel and has a hole in the center that fits the output shaft of your gear head. The output shaft of your motor will have a square notch cut into it along its length. Your wheel hub also needs a corresponding notch cut into it. When you slide the hub over the shaft, the notches line up to form a square shaft called a keyway. You can

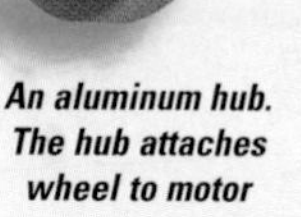

An aluminum hub. The hub attaches wheel to motor

As a rule of thumb, your weapon will take up about 20 percent of your robot's weight.

purchase a length of key stock—basically, a square metal bar—which slides into the keyway, locking the shaft and hub together. If you look around hard enough, you might even find wheels with a keyed hub already in them that will slip right over your motor's output shaft.

Photo courtesy of Richard Chandler, TeamCritter

Measuring for a keyway

The final part is the shaft collar, a little donut of steel or aluminum that clamps around the shaft and keeps the wheel from slipping off the axle.

All this machining work might be beyond the knowledge and resources of a first-time robot builder. Don't despair. There are a lot of machinists and craftsmen out there who know how to run a mill and a lathe, but don't know the first thing about radio control systems, speed controllers, motors, or batteries. If you take the time to learn how to put together these internal systems, you might find a shop or a machinist willing to join your team and contribute time and skills in exchange for the fun of helping to build a BattleBot.

If you want to make it really easy, you can purchase kits on the Web that provide a matched gear head motor, hub, and wheel. National Power Chair (www.npcinc.com) sells this type of kit, specifically for robots.

Photo courtesy of Jamie Price, Team Shenanigans

Ready to roll

All right. The motors are connected to the frame, the hubs are locked onto the motor and gear box output shafts. Now bolt the wheels onto the hubs and you've got a real, honest-to-gosh robot. You can set it on the floor and drive it around a little bit, if you promise to be careful.

Tweaking

Now that your bot is running around on its own, you can play with the radio settings. The better radios are tunable, with different response curves to choose from. You might, for example, want to make the robot's steering less sensitive at lower input levels (moving the stick a little bit) and more sensitive at higher input levels (moving the stick a lot). Or the other way around.

Photo courtesy of Richard Chandler, Team Mauser

Machining a keyway into the shaft

Many first-time builders go into the arena with a robot that over-controls. They are constantly turning too far, or going too fast, to attack the other robot effectively. You want your robot to respond quickly, but not more quickly than you can control. Experiment. Practice. Find out what works for you.

Adding the Extras

While you've got your robot running around, drive it up onto your scales and weigh it. Your finished frame and drive train should weigh in at around two-thirds of your total weight. If you're working on a Lightweight, that's about 40 pounds, or 80 pounds for a Middleweight.

Whatever weight you have left can go toward your weapon and your armor.

Photo courtesy of John Muchow, Team Dream

Now all you need is a frame to bolt to

Your Weapon

As a rule of thumb, your weapon will take up about 20 percent of your robot's weight. If you're not putting a weapon on this bot—if you're making a Wedge or a RamBot, where the body is the weapon—you can go with bigger motors and bigger batteries, and your finished drive train might be four-fifths of your robot's total weight. If, on the other hand, you're planning on packing a really big weapon, you might need to have as much as half of the robot's weight left to work with.

Christian recommends that you keep your weapon and drive systems as separate as possible, for reliability's sake. The weapons on many of his robots have their own separate batteries, RC receiver, and receiver battery.

If you go with a separate radio receiver for your weapon, you can use two radios on two different frequencies, and two operators—one to drive and one to operate the weapon. Or, if you have a 6- or 8-channel radio, you can use a single radio transmitter and two different receivers operating on the same frequency. You might use channels 1 and 2 for driving and channels 5 and 6 for the weapons, for instance.

Armor

Strictly speaking, you don't have to armor your robot, but it's a good idea.

Titanium is the preferred armor of bot builders with unlimited budgets. Tough and light, virtually Kill Saw-proof, titanium is also expensive and difficult to work with.

Aluminum is the most popular armor in the BattleBox. It's tough, light, and much cheaper than titanium. You can cut it yourself with a band saw. For armor, use the stiffer 6061 aluminum rather than the bendable 5052 aluminum.

Steel is tougher and easier to weld than aluminum, but it's also heavier.

Polycarbonate (such as Lexan) is a great material. It's more expensive than aluminum, but not much. You can paint it on the inside—it's transparent—to keep your bot looking good. You can cut it yourself or find a laser cutting house to cut out your designs and mail them to you. Another benefit of polycarbonate is that it's transparent to radio waves. You can keep your antenna protected inside your armor and it can still "see" your transmitter.

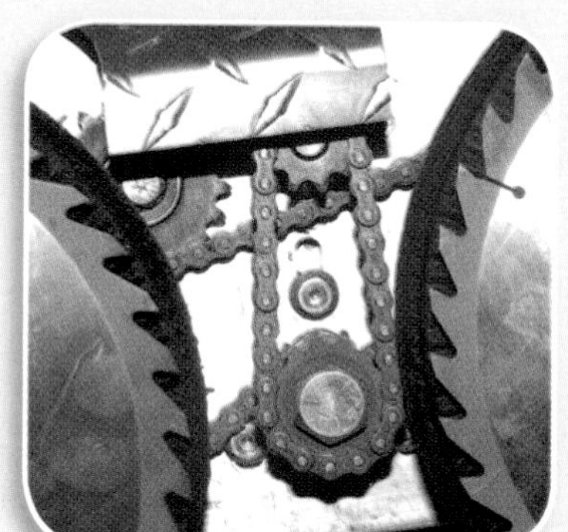

Photo courtesy of David P. Gilson, Team Critter

Chain drive

Huzzah!

That's everything. You're done. Make sure everything's locked down, all bolts and screws are snug, all batteries, electronics, and other components are securely mounted. Make sure your wire connections are good and your batteries are charged. Bolt the armor on and close the robot up.

Fire it up. Drive it around. Scare the dog. Tear up some stuff. Get to know your robot. There's no way to know exactly what a robot will do, how it will act, until you actually drive it.

When you're done admiring your new robot, it's time to look for its weaknesses. It's almost impossible to prepare for the sheer destructiveness of the BattleBox arena. As a rule, anything that can be destroyed, will be destroyed. How do you test a BattleBot for weaknesses, other than beating it with a sledgehammer? One time-honored method is to repeatedly slam your robot into the curb until something breaks. Then fix it, and slam it into the curb until something else breaks. Beating it with a sledgehammer works too, though.

> *"When you're a first-time builder, the deck is stacked against you. You've never done this before. But the sooner you start, the sooner you will figure it all out, and the more fun you will have."*
> *— Christian Carlberg*

In between beatings, practice, practice, practice. First-time builders often arrive at their first match with far too little practice, and it shows. Spend as much time as you can practicing operating your robot. Learn to stop on a dime, to turn in a flash, to bring your weapons to bear, and to dart away from trouble.

"When you're a first-time builder," says Christian, "the deck is stacked against you. You've never done this before. You don't really know what you're doing or what you're going to face in the arena. But the sooner you start, the sooner you will figure it all out, and the more fun you will have. And you can quote me on that!"

Photo courtesy of Kevin J. Dalquist

This is why solid or foam filled tires are de rigueur in BattleBots

Useful Web Links for the BattleBot Builder

The official BattleBots Web site. Tips from builders, hundreds of BattleBots, and official rules and regulations.
http://www.battlebots.com/

Coolrobots Web site. Robot links, robot database, and a step-by-step guide to building your own BattleBot.
http://www.coolrobots.com/

Team Barracuda's handy links for the first-time builder.
http://chicksdigbattlebots.botic.com/

Carlo Bertocchini's Web site features robot books, robot kits, robot parts, and robot links.
http://www.robotbooks.com/robot-design-tips.htm

Team Blaze. Combat Robot Building tips and links.
http://www.robojoust.com/

Tony Buchignani's Tips for Robot Warriors.
http://www.legalword.com/tips.html

Team Delta. Top Ten first-time builders' questions answered.
http://www.teamdelta.com/topten.htm

Team Deranged. Hints and tips for the builder.
http://members.aol.com/grayson314/hintsold.htm

Chris Hillman's Robotics. A stupefying number of robot links.
http://members.aol.com/c40179/index.html

Construction tips from the Lenox High School Bot Club.
http://www.loganbot.com/bot_tips.html

Puppetmaster Robotics. Robot builder faq.
http://puppetmaster-robotics.com/faq.html

Robot builder tips, hints, and tutorials.
http://www.teamsaber.com/

Robot Café. A great source of links, information, and discussion about robots and robot combat.
http://www.robotcafe.com/

Rockitz.com. Links and BattleBot-building tutorial.
http://www.rockitz.com/Battlebots/Building.html

Team Secret Weapon. Robot building tips and links.
http://www.geocities.com/info2x/links/links.htm

Team S.L.A.M. How to build a BattleBot.
http://users.intercomm.com/stevenn/kiss/builder.htm

Jim Smentowski's Web site. Tons of useful data, tips, faqs, advice, and history of the sport of robot combat.
http://www.robotcombat.com/

A technical guide to building fighting robots.
http://homepages.which.net/~paul.hills/index.html

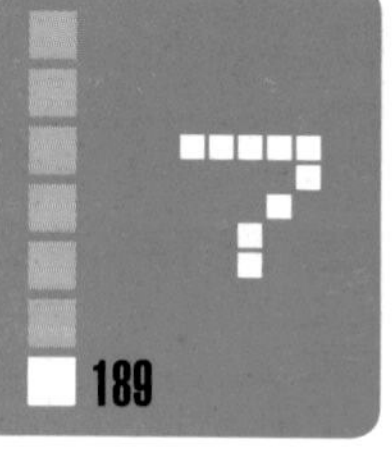

THE WINNERS

▲ Long Beach, California

April 14–15, 1999: Premiere Event

Double elimination tournament

Back then, the weight classes were designated as KiloBots (Lightweight), MegaBots (Middleweight), and GigaBots (Heavyweight). One-on-one fights were known as Robot Duels.

▲ Best of Show Award

Category	BattleBot	Team Name
Coolest Robot	Mechadon	Team Sinister
Best Engineering	BioHazard	Team BioHazard
Best Driver	KillerHurtz	Team KillerHurtz
Most Aggressive	Nightmare	Team Nightmare
Funniest Robot	Tentoumushi	Robot Action League

▲ Robot Duel Winners

KiloBots (Lightweight)

Category	BattleBot	Team Name
First Place	Ziggo	Team Ziggy
Second Place	Defiant	Team Defiant
Third Place	Executioner	Team SMC
Fourth Place	Toe Crusher	Team Coolrobots
Quarterfinalists	Crusher	Team Attitude
	Deathtrap	Team Vicious
	Sallad	Team Sallad
	Tentoumushi	Robot Action League

Zach Bieber

Christian Carlberg and Team Coolrobots

Mark Setrakian

Derek Young

MegaBots (Middleweight)

Category	BattleBot	Team Name
First Place	Son of Smashy	Automatum Technologies
Second Place	Knee Breaker	Team Coolrobots
Third place	Deadblow	Team Deadblow
Fourth Place	Turtle Road Kill	Team Attitude
Quarterfinalists	Anklebiter	Team Odin
	Carnivore	Team Carnivore
	Junior	Team Nightmare
	Stuffie	Team Stuffie

GigaBots (Heavyweight)

Category	BattleBot	Team Name
First Place	BioHazard	Team BioHazard
Second Place	KillerHurtz	Team KillerHurtz
Third Place	Rhino	Inertia Labs
Fourth Place	Tazbot	Mutant Robots
Quarterfinalists	Nightmare	Team Nightmare
	Punjar	Team Punjar
	Razer	Team Razer
	Vlad the Impaler	Vladmeisters

▲ Robot Rumble Winners

Category	BattleBot	Team Name
KiloBots (LW)	Hammerhead	Team Brute Force
MegaBots (MW)	Deadblow	Team Deadblow
GigaBots (HW)	Razer	Team Razer

Las Vegas, Nevada

November 17, 1999: Pay-Per-View

Single elimination tournament

The metric/hard drive designation was eliminated for the weight classes and standard designations—Lightweight, Heavyweight, etc.—were used. This was the first appearance of the Super Heavyweight class. Since this was pay-per-view and there was limited time, BattleBots opted to have a Heavyweight and Super Heavyweight tournament only.

Best of Show Award

Category	BattleBot	Team Name
Coolest Robot	Mechadon	Team Sinister
Best Engineering	Mechadon	Team Sinister
Best Driver	Punjar	Team Punjar
Most Aggressive	Mortis	Random Violence Technology

Robot Duel Winners

Heavyweight

Category	BattleBot	Team Name
First Place	Vlad the Impaler	Vladmeisters
Second Place	Voltarc *(now Voltronic)*	Team Voltronic
Semifinalists	Punjar	Team Punjar
	Rhino	Inertia Labs
Quarterfinalists	BioHazard	Team BioHazard
	KillerHurtz	Team KillerHurtz
	Mortis	Random Violence Technology
	Nightmare	Team Nightmare

Deadblow carries the Nut

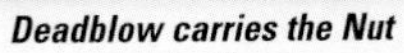

Jim Smentowski wins

Super Heavyweight

Category	BattleBot	Team Name
First Place	Minion	Team Coolrobots
Second Place	Ricon *(now Odin II)*	Team Odin
Semifinalists	Mechadon	Team Sinister
	World Peace	Team Bohica
Quarterfinalists	Abattoir	Team Wetware
	DooAll	Team DooAll
	Ginsu	La Ma Motors
	S.L.A.M.	Team S.L.A.M.

Robot Rumble Winners

Category	BattleBot	Team Name
Heavyweight	BioHazard	Team BioHazard
Super Heavyweight	World Peace	Team Bohica

John Ridder

Klotonya Hamilton

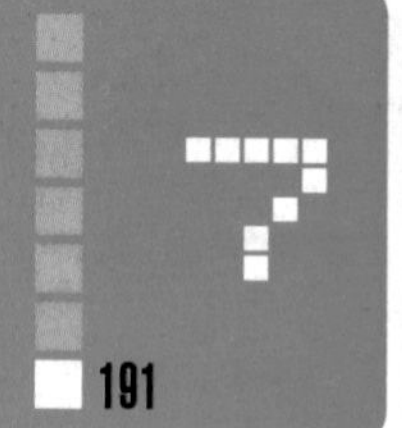

San Francisco, California

June 9–11, 2000: Comedy Central Season 1.0

Single elimination tournament

This was the first show with Comedy Central. It was an all-weight-class event.

Best of Show Award

Category	BattleBot	Team Name
Coolest Robot	Mechadon	Team Sinister
Best Newcomer *(tie)*	Suicidal Tendencies	Team Suicide
	Alpha Raptor *(now Gamma Raptor)*	Team Raptor
Best Engineering	Mechadon	Team Sinister
Best Driver	BioHazard	Team BioHazard
Most Aggressive	Mauler	South Bay RoboWarriers

Robot Duel Winners

Lightweight

Category	BattleBot	Team Name
First Place	Backlash	Team Nightmare
Second Place	Alpha Raptor	Team Raptor
Semifinalists	Das Bot	Team Das Bot
	Mouser Mecha-Catbot	Team Catbot
Quarterfinalists	Endotherm	Team Saber
	No Tolerance III	No Tolerance Combat Robotics
	Tentoumushi	Robot Action League
	Ziggo	Team Ziggy

Middleweight

Category	BattleBot	Team Name
First Place	Hazard	Team Hazard
Second Place	Deadblow	Team Deadblow
Semifinalists	Pressure Drop	Automatum Technologies
	Super Orbiting Force	Robot Action League
Quarterfinalists	Alien Gladiator	Team Malicious
	Anklebiter	Team Odin
	Blade Runner	Team Carnivore
	Turtle Road Kill	Team Attitude

Heavyweight

Category	BattleBot	Team Name
First Place	Vlad the Impaler	Vladmeisters
Second Place	Voltarc *(now Voltronic)*	Team Voltronic
Semifinalists	KillerHurtz	Team KillerHurtz
	Punjar	Team Punjar
Quarterfinalists	BioHazard	Team BioHazard
	Gammatron	Gammatronic Robot Brigade
	Mauler	South Bay RoboWarriors
	Overkill	Team Coolrobots

Super Heavyweight

Category	BattleBot	Team Name
First Place	Minion	Team Coolrobots
Second Place	DooAll	Team DooAll
Semifinalists	Rammstein	Loki Robotics
	Ronin	Team Sinister
Quarterfinalists	Ginsu	La Ma Motors
	Grendel *(now Revision Z)*	Team Malicious
	Rhino	Inertia Labs
	Mechadon	Team Sinister

Robot Rumble Winners

Category	BattleBot	Team Name
Lightweight	Mouser Mecha-Catbot	Team Catbot
Middleweight	Hazard	Team Hazard
Heavyweight	BioHazard	Team BioHazard
Super Heavyweight	Minion	Team Coolrobots

Thomas Peteruccelli wins

Grant Imahara

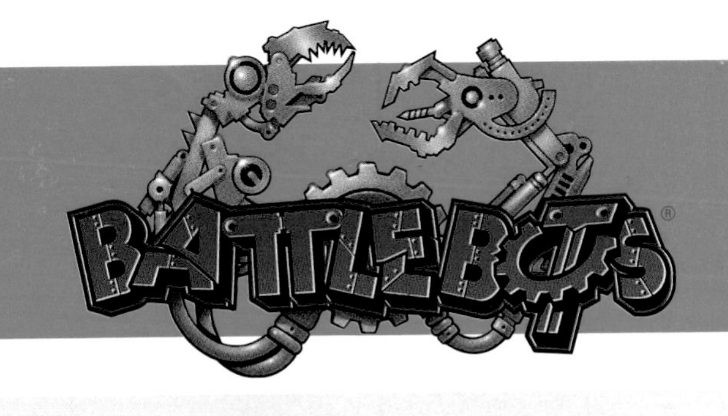

▲ Las Vegas, Nevada

Nov 16–19, 2000: Comedy Central Season 2.0

Single elimination tournament

This was the second show with Comedy Central, another all-weight-class event.

▲ Best of Show Award

Category	BattleBot	Team Name
Coolest Robot	Snake	Team Sinister
Best Engineering	Snake	Team Sinister
Most Aggressive	Ziggo	Team Ziggy

▲ Robot Duel Winners

Lightweight

Category	BattleBot	Team Name
First Place	Ziggo	Team Ziggy
Second Place	Backlash	Team Nightmare
Semifinalists	Beta Raptor *(now Gamma Raptor)*	Team Raptor
	Toe Crusher	Team Coolrobots
Quarterfinalists	Afterthought	Robot Action League
	Crusher	Team Attitude
	Evil Fish Tank	Team Delta
	Sallad	Team Sallad

Middleweight

Category	BattleBot	Team Name
First Place	Spaz	Team Vicious
Second Place	El Diablo	Team Diablo
Semifinalists	Bad Attitude	Team Attitude
	The Master	Team Sinister
Quarterfinalists	Blade Runner	Team Carnivore
	Buddy Lee Don't Play in the Street	Team Fembot
	Complete Control	Automatum Technologies
	Deadblow	Team Deadblow

Heavyweight

Category	BattleBot	Team Name
First Place	BioHazard	Team BioHazard
Second Place	Vlad the Impaler	Vladmeisters
Semifinalists	FrenZy	Team Minus Zero
	Voltronic	Team Voltronic
Quarterfinalists	FrostBite	Team Toad
	GoldDigger	Team Boomer
	Nightmare	Team Nightmare
	Tazbot	Mutant Robots

Super Heavyweight

Category	BattleBot	Team Name
First Place	Diesector	Mutant Robots
Second Place	Atomic Wedgie	Team Half-Life
Semifinalists	Revision Z	Team Malicious
	War Machine	Team Delta
Quarterfinalists	DooAll	Team DooAll
	Rammstein	Loki Robotics
	Ronin	Team Sinister
	Toro	Inertia Labs

▲ Robot Rumble Winners

Category	BattleBot	Team Name
Lightweight	HammerHead	Team Brute Force
Middleweight	Deadblow	Team Deadblow
Heavyweight	Tazbot	Mutant Robots
Super Heavyweight	Toro	Inertia Labs

Team Fembot

Ian Lewis and Simon Scott

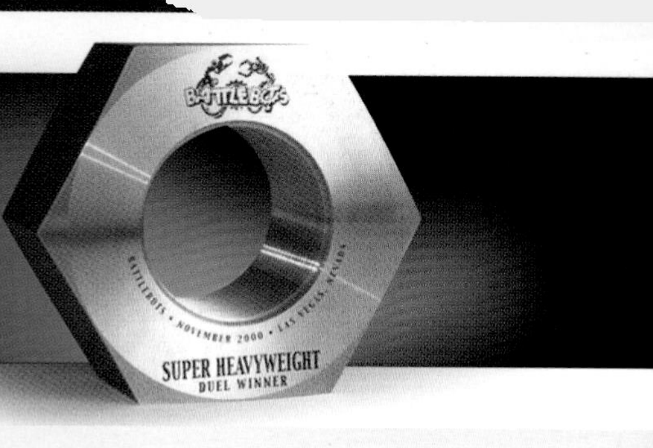

San Francisco, California

May 22–28, 2001: Comedy Central Season 3.0

Single elimination tournament

This all-weight-class event was the third show with Comedy Central.

▲ Best of Show Award

Category	BattleBot	Team Name
Coolest Robot	T-Minus	Inertia Labs
Best Engineering	The Judge	Mechanicus
Most Aggressive	The Judge	Mechanicus

▲ Robot Duel Winners

Lightweight

Category	BattleBot	Team Name
First Place	Dr. Inferno Jr.	The Infernolab
Second Place	Gamma Raptor	Team Raptor
Semifinalists	Sallad	Team Sallad
	Wedge of Doom	Team Hazard
Quarterfinalists	Backlash	Team Nightmare
	Herr Gepoünden	Team Zwölfpack XII
	Mouser Mecha-Catbot	Team Catbot
	Ziggo	Team Ziggy

Middleweight

Category	BattleBot	Team Name
First Place	Hazard	Team Hazard
Second Place	Little Drummer Boy	BotWorks *(now Team Dangerous Drum)*
Semifinalists	SABotage	Team Sabotage
	T-Wrex	Regan Designs
Quarterfinalists	Eraser	Pencil Neck Geeks
	F5	Unidentified Flying Object
	T-Minus	Inertia Labs
	Twin Paradox	Skunk-Tek

Heavyweight

Category	BattleBot	Team Name
First Place	Son of Whyachi	Team Whyachi
Second Place	BioHazard	Team BioHazard
Semifinalists	Hexadecimator	Team WhoopAss
	Overkill	Team Coolrobots
Quarterfinalists	KillerHurtz	Team KillerHurtz
	MechaVore	Team Shrapnel
	Tazbot	Mutant Robots
	Voltronic	Team Voltronic

Super Heavyweight

Category	BattleBot	Team Name
First Place	Vladiator	Vladmeisters
Second Place	Minion	Team Coolrobots
Semifinalists	Diesector	Mutant Robots
	Toro	Inertia Labs
Quarterfinalists	Electric Lunch	Team K.I.S.S.
	Phere	Team Xtremebots
	Rammstein	Loki Robotics
	Techno Destructo	Team Carnage

▲ Robot Rumble Winners

Category	BattleBot	Team Name
Lightweight	Ziggo	Team Ziggy
Middleweight	T-Minus	Inertia Labs
Heavyweight	Hexadecimator	Team WhoopAss
Super Heavyweight	Toro	Inertia Labs

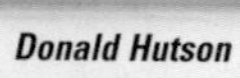
Donald Hutson

Carlo Bertocchini

San Francisco, California

Nov 4–11, 2001: Comedy Central Season 4.0

Single elimination tournament

This was the fourth show with Comedy Central, another all-weight-class event. BattleBots changed the name Robot Duel to BotBout.

▲ Best of Show Award

Category	BattleBot	Team Name
Coolest Robot *(tie)*	Tazbot	Mutant Robots
	Toro	Inertia Labs
Best Engineering	The Judge	Mechanicus
Best Driver	Donald Hutson, *for Tazbot and Diesector*	Mutant Robots
Most Aggressive	M.O.E. *(Marvel of Engineering)*	Team Flaming Monkeys

▲ BotBout Winners

Lightweight

Category	BattleBot	Team Name
First Place	Ziggo	Team Ziggy
Second Place	The Big B	Late Night Racing
Semifinalists	Carnage Raptor	Team Raptor
	Death by Monkeys	Team Death by Monkeys
Quarterfinalists	Hexy Jr.	Team WhoopAss Jr.
	Sallad	Team Sallad
	Slap 'Em Silly	Fatcats
	Wedge of Doom	Team Hazard

Middleweight

Category	BattleBot	Team Name
First Place	Hazard	Team Hazard
Second Place	Complete Control	Automatum Technologies
Semifinalists	Heavy Metal Noise	Big Bang Robotics
	Zion	Zion
Quarterfinalists	Bad Attitude	Team Attitude
	El Diablo	Team Diablo
	Huggy Bear	Team Huggy Bear
	Twin Paradox	Skunk-Tek

Heavyweight

Category	BattleBot	Team Name
First Place	BioHazard	Team BioHazard
Second Place	Overkill	Team Coolrobots
Semifinalists	Surgeon General	Loki Robotics
	Tazbot	Mutant Robots
Quarterfinalists	Hexadecimator	Team WhoopAss
	M.O.E. *(Marvel of Engineering)*	Team Flaming Monkeys
	Nightmare	Team Nightmare
	Slam Job	Blackroot

Super Heavyweight

Category	BattleBot	Team Name
First Place	Toro	Inertia Labs
Second Place	New Cruelty	Killerbotics
Semifinalists	Diesector	Mutant Robots
	Little Blue Engine	Team Circuit Breaker
Quarterfinalists	Iceberg	Team Toad
	Swirlee	Pain USA
	Techno Destructo	Team Carnage
	Vladiator	Vladmeisters

▲ Robot Rumble Winners

Category	BattleBot	Team Name
Lightweight	Dr. Inferno Jr.	The Infernolab
Middleweight	Bad Attitude	Team Attitude
Heavyweight	Little Sister	Team Big Brother
Super Heavyweight	The Judge	Mechanicus

BATTLESTATS

Robot Name	Team	Ranking	Weapon	W/L (W/L%)
2925 Jack Rabbit	Team Brown	#19 Middleweight	Custom fang	5/1 (83.33%)
401(k)		#141 Lightweight	Elastomer robot body	0/1 (0.00%)
A Bot Named Sue	A Team in Black	#150 Lightweight	Carbide-tipped steel blades; steel-spiked wheels	0/1 (0.00%)
A.W.O.L	Team Black Op Six	#136 Lightweight	Buzz saw	0/2 (0.00%)
Abaddon	Team Shoggoth	#37 Super Heavyweight	2 horizontal hammers; flopping wedge	2/1 (66.67%)
Abattoir	Team Wetware	#85 Super Heavyweight	Counter-rotating wheel blades	0/4 (0.00%)
ACK ACK	Truly Unruly Robotics	#112 Lightweight		1/1 (50.00%)
ACME Surplus Fighting Robot		#168 Lightweight	Dual 5-in. spinning drums	0/1 (0.00%)
The Administrator	Edge Robotics	#88 Heavyweight		0/1 (0.00%)
Adult Supervision Is Required	General Technics	#160 Middleweight	2 long lifting forks	0/1 (0.00%)
ÆON 00	Team Big Fire	#169 Lightweight	Battering ram	0/1 (0.00%)
Afterburner		#46 Lightweight	Spinning shell with two hardened S7 tool steel knockers	2/1 (66.67%)
Aftermath	Team Black Op Six	#151 Middleweight	Pike of Doom and Weapon Pod	0/2 (0.00%)
Afterthought	Robot Action League	#26 Lightweight	18-in. 2500 RPM steel disk with cutter blades	4/3 (57.14%)
Aggravated Assault	Real Good Robotics	#86 Lightweight	Bulldozer blade	1/1 (50.00%)
The Aggressive Polygon	Robot Action League	#163 Lightweight		0/1 (0.00%)
Agitator	Robotdojo	#48 Heavyweight	2 spinning arms	2/1 (66.67%)
Agrippa	Team Delta	#80 Heavyweight		1/2 (33.33%)
Agsma	Neilson	#117 Middleweight	Spinning blade	0/1 (0.00%)
Alabama Slammer	Redneck Robots	#26 Middleweight	Pitchfork flipper	4/2 (66.67%)
Alcoholic Stepfather	Team Kahuna	#72 Super Heavyweight	2, 6-pound sledgehammers	0/1 (0.00%)
Alien 2	Team Malicious	#171 Lightweight		0/1 (0.00%)
Alien Gladiator	Team Malicious	#105 Middleweight	Spring-loaded sledgehammer	1/1 (50.00%)
All Weather Tire Bot	knetx.com	#142 Middleweight	Auto tire with snow chains	0/1 (0.00%)
Alpha Crusader		#145 Lightweight	Vertical whacker with spikes	0/1 (0.00%)
American Justice	Team BattleRat	#56 Super Heavyweight	Dinosaur-like body-piercing arches	1/1 (50.00%)
Angriff	Five-O-Five	#83 Super Heavyweight	Lifting/ramming spike	0/1 (0.00%)
Anklebiter	Team Odin	#72 Middleweight		2/6 (25.00%)
Annihilator	Team Brute Force	#101 Heavyweight	Lawnmower blade	0/1 (0.00%)
Annuit Geptus	Annuit Coeptus	#204 Lightweight		0/2 (0.00%)
The Annihilator	The Bulldogs	#126 Middleweight		0/1 (0.00%)
Anubis	Hunter Robotics	#57 Super Heavyweight	Lifter arms; 8-in. carbide-tipped circular saw blade	1/2 (33.33%)
Archangel	Genesis Robotics	#137 Lightweight	Long, movable arm with saw blade mounted at end	0/1 (0.00%)

LEGEND

W/L (W/L%) : Wins/Losses (%)
TT : Total Tournaments
TM : Total Matches
KOs : Knock Outs
KO% : Knock Out Percentage
AKOT : Average KO Time
KOA : Knock Out Against
KOA% : KO Against Percentage
AKOAT : Average KO Against Time
TJD : Total Judge Decisions
JPA : Judge Point Average
AA : Aggression Average
DA : Damage Average
SA : Strategy Average

TT	TM	KOs	KO%	AKOT	KOA	KOA%	AKOAT	TJD	JPA	AA	DA	SA	Points	Builder	Robot Name
1	6	1	16.66	0:05:38	0	0	0:00:00	3	25.33	9	7.67	8.67	10	TJ (Terrence) Brown	2925 Jack Rabbit
1	1	0	0	0:00:00	0	0	0:00:00	1	14	3	4	7	0	David Whitaker	401(k)
1	1	0	0	0:00:00	0	0	0:00:00	1	7	2	4	1	0	Christopher Powell	A Bot Named Sue
2	2	0	0	0:00:00	1	50	0:01:50	1	16	5	4	7	0	Steven West	A.W.O.L.
1	3	1	33.33	0:02:38	0	0	0:00:00	2	29	10.5	10.5	8	4	Kenneth Converse	Abaddon
4	3	0	0	0:00:00	2	66.66	0:01:31	1	1	0	0	1	0	Ray Scully	Abattoir
1	2	0	0	0:00:00	1	50	0:00:57	0	0	0	0	0	2	John Waldron	ACK ACK
1	1	0	0	0:00:00	1	100	0:01:27	0	0	0	0	0	0	Jason Demerski	ACME Surplus Fighting Robot
1	1	0	0	0:00:00	0	0	0:00:00	1	18	6	6	6	0	Austin Carpenter	Administrator, The
1	1	0	0	0:00:00	0	0	0:00:00	0	0	0	0	0	0	Robert Wenzlaff	Adult Supervision Is Required
1	1	0	0	0:00:00	1	100	0:01:10	0	0	0	0	0	0		ÆON 00
1	3	2	66.66	0:00:56	0	0	0:00:00	1	21	4	7	10	4	Nola Garcia	Afterburner
2	2	0	0	0:00:00	1	50	0:02:18	0	0	0	0	0	0	Steven West	Aftermath
3	7	2	28.57	0:02:32	1	14.28	0:01:55	3	10.33	2.67	5	2.67	8	Jim Sellers	Afterthought
1	2	1	50	0:01:47	1	50	0:01:36	0	0	0	0	0	2	Rob Gibson-Dugger	Aggravated Assault
1	1	0	0	0:00:00	0	0	0:00:00	1	0	0	0	0	0	Will Wright	Aggressive Polygon, The
1	3	0	0	0:00:00	0	0	0:00:00	2	28	10	10	8	4	Mike Konshak	Agitator
1	3	0	0	0:00:00	1	33.33	0:02:45	0	0	0	0	0	2	Dan Danknick	Agrippa
1	1	0	0	0:00:00	0	0	0:00:00	1	19	6	6	7	0	John Neilson	Agsma
2	6	1	16.66	0:00:31	2	33.33	0:00:47	3	35.67	11.67	12.33	11.67	8	Richard Nickel	Alabama Slammer
1	1	0	0	0:00:00	0	0	0:00:00	1	19	6	11	2	0	Zain Saidin	Alcoholic Stepfather
1	1	0	0	0:00:00	1	100	0:02:10	0	0	0	0	0	0	Daniel Rupert	Alien 2
1	2	0	0	0:00:00	1	50	0:01:08	0	0	0	0	0	2	Daniel Rupert	Alien Gladiator
1	1	0	0	0:00:00	0	0	0:00:00	1	7	2	3	2	0	Travis Wood	All Weather Tire Bot
1	1	0	0	0:00:00	0	0	0:00:00	1	11	4	4	3	0	Michael Obrien	Alpha Crusader
1	2	0	0	0:00:00	0	0	0:00:00	2	23.5	8.5	6.5	8.5	2	Ted Walters	American Justice
1	1	0	0	0:00:00	0	0	0:00:00	1	6	1	2	3	0	Rick Perotti	Angriff
5	8	1	12.5	0:02:22	5	62.5	0:02:25	0	0	0	0	0	3	John McKenzie	Anklebiter
1	1	0	0	0:00:00	0	0	0:00:00	1	5	5	0	0	0	Charles Steinkuehler	Annihilator
1	0	0	0	0:00:00	0	0	0:00:00	0	0	0	0	0	0	Jeremy Rutman	Annuit Geptus
1	1	0	0	0:00:00	0	0	0:00:00	1	13	4	3	6	0	Damian Quintana	Annihilator, The
2	3	0	0	0:00:00	1	33.33	0:02:12	2	21	6.5	8.5	6	2	Glen Sears	Anubis
1	1	0	0	0:00:00	0	0	0:00:00	1	15	3	6	6	0	Antoine Trabulsi	Archangel

Robot Name	Team	Ranking	Weapon	W/L (W/L%)
The Archduke	Team Poison Fists	#44 Lightweight		3/1 (75.00%)
Aries	Odyssey	#172 Lightweight	Steel ramming blade	0/1 (0.00%)
Armadillo	Creative Anarchy	#35 Middleweight	Ramming	3/2 (60.00%)
Armageddon	Mayhem	#39 Super Heavyweight	Wedge	2/2 (50.00%)
Atomic Wedgie	Team Half-Life Inc.	#12 Super Heavyweight	Conveyor belts; spinning tri-foils	4/3 (57.14%)
Avatar	Project X	#162 Middleweight	Saws; spikes	0/1 (0.00%)
B.E.H.EMOTH	Team B.E.H.EMOTH	#78 Middleweight	2 pneumatic flippers; twin buzz saws	1/1 (50.00%)
B.O.B. (Basher of BOTS)	Team B.O.B.	#42 Heavyweight	The Disc of Death (B.O.B. the Ripper)	2/1 (66.67%)
Bacchus	Lafayette	#17 Heavyweight	Flipping arm	4/1 (80.00%)
Backlash	Team Nightmare	#2 Lightweight	Spinning disk	12/3 (80.00%)
Bad Attitude	Team Attitude	#7 Middleweight	Wedge	6/4 (60.00%)
Bad Badger 2.0	Team S.M.A.C.	#58 Heavyweight	Linear actuator-driven claws; steel bumper	1/1 (50.00%)
Bad Bone	Team Bone	#173 Lightweight	Lifting arm	0/1 (0.00%)
Bad Dog	K-NINE	#54 Heavyweight	Wedge	1/1 (50.00%)
Bad Habit	Team Widget	#59 Lightweight	Wedge	2/2 (50.00%)
Bad News		#106 Lightweight	1500 RPM spinning drum	1/1 (50.00%)
Bait	People in the Basement	#62 Lightweight	4-in. grappling hook on a long pole	2/1 (66.67%)
Ballbreaker	Balls of Steel	#93 Super Heavyweight	Angular momentum	0/1 (0.00%)
Bambino	Bambino	#94 Lightweight	Baseball bats	1/2 (33.33%)
Bang	Team Bang	#163 Middleweight	Spiked tail; spinning saw blade	0/1 (0.00%)
Batray	Team Think Tank	#200 Middleweight	Four corners (carbide or high-speed steel) of the bot	0/1 (0.00%)
Batrok Z	Team Skuld	#164 Middleweight	Pneumatic steel spike	0/1 (0.00%)
Battle Scar	MoBot	#74 Super Heavyweight	Steel saw blade	0/1 (0.00%)
The BattleRat	Team BattleRat	#27 Heavyweight	Lifting arm; wedge	3/2 (60.00%)
Bender	Tatar	#56 Heavyweight	60-lb. 2000 RPM spinning blade	1/4 (20.00%)
Berzerker 2000	O U Death Squad	#135 Heavyweight	36-inch diameter armored tire, with interchangeable weapons	0/1 (0.00%)
The Bevo Blender	Sooners	#97 Heavyweight	Locked flying wedge	0/1 (0.00%)
The Big B	Late Night Racing	#6 Lightweight	Front and back wedges	8/1 (88.89%)
Big Betty	Axonn	#136 Heavyweight		0/1 (0.00%)
Big Stick	The Doom Crew	#174 Lightweight	2-ft. pivoting blades	0/1 (0.00%)
Bigger Brother	Team Big Brother	#61 Heavyweight	CO_2-powered flipper	1/1 (50.00%)
Billy-Bot	TrailerTrash	#78 Heavyweight	24-in. circular saw; 10-in. front-mounted spike	1/1 (50.00%)
BioHazard	Team BioHazard	#1 Heavyweight	Lifting arm	22/3 (88.00%)
The Black Knight	The Covina Guys	#94 Super Heavyweight	2, 16-lb. rotating sledgehammers	0/1 (0.00%)
Black Ops	Black Ops	#104 Heavyweight	Deployable arms	0/2 (0.00%)
Black Warrior	Team Nose Probe	#82 Lightweight	Wedge	1/1 (50.00%)
Black Widow	Team Black Widow	#175 Lightweight	Kinetic energy	0/1 (0.00%)
Blade Runner	Team Carnivore	#25 Middleweight	35-lb. pickax; ramming spikes	4/4 (50.00%)
Blendo	Team Blendo	#103 Heavyweight	Spinner	0/4 (0.00%)
Blood Dragon	Raving Lunatics	#66 Lightweight	Arm with armored spike, scoop, or hook	2/1 (66.67%)

TT	TM	KOs	KO%	AKOT	KOA	KOA%	AKOAT	TJD	JPA	AA	DA	SA	Points	Builder	Robot Name
1	4	0	0	0:00:00	0	0	0:00:00	3	23.67	7.67	9.33	6.67	6	John Howard	Archduke, The
1	1	0	0	0:00:00	1	100	0:00:49	0	0	0	0	0	0	John Crain	Aries
2	5	1	20	0:00:43	1	20	0:00:56	3	29.67	10	9.67	10	6	Mike Kalkwarf	Armadillo
2	4	0	0	0:00:00	0	0	0:00:00	3	26.33	9.33	8.33	8.67	4	Doug Groves	Armageddon
3	7	2	28.57	0:01:51	0	0	0:00:00	5	20.8	7	7.4	6.4	9	Robert Everhart	Atomic Wedgie
1	1	0	0	0:00:00	1	100	0:00:34	0	0	0	0	0	0	Joesph Butler	Avatar
1	2	1	50	0:01:16	0	0	0:00:00	1	17	6	5	6	2	Eric Hilgeford	B.E.H.EMOTH
1	3	1	33.33	0:01:39	0	0	0:00:00	2	19.5	6.5	6.5	6.5	4	Wray Russ	B.O.B. (Basher of BOTS)
1	5		40	0:01:41	0	0	0:00:00	3	30.33	9.33	11.33	9.67	8	Lafayette College	Bacchus
4	15	10	66.66	0:01:33	3	20	0:01:00	2	39	12	12.5	14.5	27	Jim Smentowski	Backlash
4	10	3	30	0:01:46	2	20	0:02:00	5	25.2	8.4	8.4	8.4	14	Thomas Petruccelli	Bad Attitude
1	2	1	50	0:01:03	0	0	0:00:00	1	3	1	0	2	2	Ryan Lufkin	Bad Badger 2.0
1	1	0	0	0:00:00	1	100	0:01:50	0	0	0	0	0	0	Will Keels	Bad Bone
1	2	1	50	0:00:43	1	50	0:01:17	0	0	0	0	0	2	Kenneth Benner	Bad Dog
2	4	1	25	0:01:40	2	50	0:01:36	1	23	9	6	8	4	Ray Alderman	Bad Habit
1	2	0	0	0:00:00	1	50	0:01:04	1	28	4	10	14	2		Bad News
1	3	0	0	0:00:00	0	0	0:00:00	2	28	10	8	10	4	Joe Ludwig	Bait
1	1	0	0	0:00:00	1	100	0:00:52	0	0	0	0	0	0	Shawn Ouderkirk	Ballbreaker
2	3	0	0	0:00:00	0	0	0:00:00	3	17.33	6.33	5.33	5.67	2	Steve Wilson	Bambino
1	1	0	0	0:00:00	1	100	0:01:11	0	0	0	0	0	0	Troy Alton	Bang
1	0	0	0	0:00:00	0	0	0:00:00	0	0	0	0	0	0	Ted Shimoda	Batray
1	1	0	0	0:00:00	1	100	0:00:37	0	0	0	0	0	0	Brett Bellmore	Batrok Z
1	1	0	0	0:00:00	0	0	0:00:00	1	18	6	6	6	0	Brad Broman	Battle Scar
2	5	1	20	0:01:24	0	0	0:00:00	4	21.5	8.25	7	6.25	6	Ted Walters	BattleRat, The
4	5	1	20	0:00:54	2	40	0:00:54	2	5	2	1	2	2	Eric Stoliker	Bender
1	0	0	0	0:00:00	0	0	0:00:00	0	0	0	0	0	0	Joey Sala	Berzerker 2000
1	1	0	0	0:00:00	0	0	0:00:00	1	8	4	3	1	0	John Fagan	Bevo Blender, The
1	9	1	11.11	0:01:34	0	0	0:00:00	7	28.71	9.86	9.57	9.29	17	Gary Gin	Big B, The
1	0	0	0	0:00:00	0	0	0:00:00	0	0	0	0	0	0	Henry L. Hillman Jr.	Big Betty
1	1	0	0	0:00:00	1	100	0:01:09	0	0	0	0	0	0	Terry Wynn	Big Stick
1	2	1	50	0:01:20	0	0	0:00:00	1	2	1	0	1	2	Ian Watts	Bigger Brother
1	2	0	0	0:00:00	0	0	0:00:00	1	10	4	5	1	0	Michael Schermer	Billy-Bot
6	25	10	40	0:01:45	1	4	0:02:13	12	30.33	10.17	10.42	9.75	55	Carlo Bertocchini	BioHazard
1	1	0	0	0:00:00	1	100	0:01:30	0	0	0	0	0	0	Matt Sandt	Black Knight, The
2	2	0	0	0:00:00	2	100	0:01:19	0	0	0	0	0	0	Rob Meyer	Black Ops
1	2	1	50	0:01:28	1	50	0:01:24	0	0	0	0	0	2	Keegan Hutto	Black Warrior
1	1	0	0	0:00:00	1	100	0:00:55	0	0	0	0	0	0	Timothy Montagne	Black Widow
4	8	3	37.5	0:01:58	2	25	0:01:49	3	27.33	10.33	6.67	10.33	8	Ilya Polyakov	Blade Runner
4	4	0	0	0:00:00	4	100	0:01:45	0	0	0	0	0	0	Jamie Hyneman	Blendo
1	3	0	0	0:00:00	0	0	0:00:00	2	17.5	6	5.5	6	4	Malcolm Clark	Blood Dragon

Robot Name	Team	Ranking	Weapon	W/L (W/L%)
Blood Moon Mark II	CACK	#42 Lightweight	Vertical spinning square	3/2 (60.00%)
Blue Streak	Team J T 2	#130 Middleweight	Lifting jaws	0/1 (0.00%)
Blunt Force Trauma	Blunt Force Trauma	#105 Heavyweight	Horizontal plane rotating inertia	0/2 (0.00%)
Bolo Mark X		#82 Super Heavyweight		0/1 (0.00%)
Bombus Bombus	Team Finagler	#205 Lightweight		0/2 (0.00%)
Booby Trap	Team Half-Life Inc.	#50 Middleweight	Pneumatic flipper; spiked hammer	2/1 (66.67%)
Bot Mulcher	Rio Grande	#165 Middleweight	Horizontal spinning blade	0/1 (0.00%)
BotABing	Cloud Cover Robotics	#152 Middleweight	Wedge and air chisel	0/2 (0.00%)
Botilla the Hun	Team Mobot	#151 Lightweight	(top secret)	0/2 (0.00%)
Botknocker 19	Team Botknocker 19	#36 Heavyweight	Rotating hull	2/2 (50.00%)
Bottom Dweller	Carter	#124 Middleweight	Lift arm	0/2 (0.00%)
Botulism	Team Insight	#97 Middleweight	Spinning 6-lb. mauls; pneumatic tail with flipper, pickax, or spear	1/1 (50.00%)
Botulizer	Killer-bot	#23 Heavyweight	Movable scoop	3/1 (75.00%)
Bouncing Betty	Axonn	#20 Middleweight	Flipping arm	5/2 (71.43%)
Brain Mold	Poor College Guys	#146 Lightweight	Lifter	0/1 (0.00%)
The Brainsters of Triskellian	Ryan Engineering	#59 Super Heavyweight		1/1 (50.00%)
Bucky the Beaverbot	TeamBeaverbot	#29 Super Heavyweight	Horizontal spinner mass; wedged snout with spike teeth	2/1 (66.67%)
Buddy Lee Don't Play in the Street	Team Fembot	#61 Middleweight	Spinning blade	2/3 (40.00%)
Bulldog	Computer Keyes	#32 Lightweight	Movable wedge/plow/lifting arm	3/2 (60.00%)
Bumble Bee	Dynamic Duo	#137 Heavyweight	Titanium blade, powered by three 6hp gasoline engines	0/1 (0.00%)
Bumper	Tilt-Con	#121 Middleweight	50-lb. 2000 RPM spinning disk	0/1 (0.00%)
BumperBawt	Bawtz	#153 Middleweight	Spinning Claw of Pain	0/2 (0.00%)
Burning Metal	Outlaw	#75 Lightweight	Horizontal spinning arm	1/1 (50.00%)
Bust-O-Matic	Bust-O-Matic	#81 Lightweight	3, 2-lb. 1000 RPM spinning sledgehammer heads	1/1 (50.00%)
Caliban	Team Whoop-Ass	#70 Middleweight	Spinning blades	2/2 (50.00%)
Cannibot	Cannibal Robotics	#75 Super Heavyweight	Pneumatic lifter arm	0/1 (0.00%)
Captain Insane-O	SDSU Aztecs	#40 Middleweight	4000 RPMs and invertible butt flap	3/1 (75.00%)
Carnage Raptor	Team Raptor	#12 Lightweight	Ax	6/2 (75.00%)
Carnivore	Team Carnivore	#80 Middleweight		1/2 (33.33%)
Center Punch	Green Iguana	#47 Heavyweight	Large steel spike	2/2 (50.00%)
Centipede	Myriapod	#130 Lightweight	3-ft. horizontal blade	0/1 (0.00%)
Cerberus	Team Hells Mechanics	#71 Middleweight	3-sided wedge	2/1 (66.67%)
Chains Addiction	Bad Seeds	#99 Heavyweight	4 chains tipped with steel blocks, spinning at 4000 RPM	0/1 (0.00%)
Chameleon	Renegade Robots	#109 Middleweight	Rear spikes; front modular rotational weapon	0/2 (0.00%)
Cheap Shot	Poor College Guys	#55 Middleweight	Pneumatic steel flipping fork	2/1 (66.67%)
Chewbot	Dreadbot	#57 Heavyweight	Spinning bar; wedge	1/1 (50.00%)
Chewy	Zombie918	#176 Lightweight	Wedge	0/1 (0.00%)
ChiaBot	Robot Action League	#177 Lightweight	Spinning Wheel of Love	0/1 (0.00%)
Chiabot	Robot Action League	#67 Middleweight	Spinning Wheel of Love	2/3 (40.00%)
Chimera	AVHS Robotics	#120 Lightweight	Wedge; CO_2 plow-lift; 6.5-in. circular saw	0/2 (0.00%)

TT	TM	KOs	KO%	AKOT	KOA	KOA%	AKOAT	TJD	JPA	AA	DA	SA	Points	Builder	Robot Name
2	5	0	0	0:00:00	2	40	0:01:36	3	28.33	9.67	10	8.67	6	Jordan Hackney	Blood Moon Mark II
1	1	0	0	0:00:00	0	0	0:00:00	1	12	4	5	3	0	Jason Tajima	Blue Streak
2	2	0	0	0:00:00	2	100	0:01:20	0	0	0	0	0	0	Kevin Walsh	Blunt Force Trauma
1	1	0	0	0:00:00	0	0	0:00:00	1	8	2	3	3	0	Carl Kalkhof	Bolo Mark X
1	0	0,	0	0:00:00	0	0	0:00:00	0	0	0	0	0	0	Jack Buffington	Bombus Bombus
1	3	1	33.33	0:00:48	0	0	0:00:00	2	28	10	9	9	4	Kimberly Everhart	Booby Trap
1	1	0	0	0:00:00	1	100	0:00:31	0	0	0	0	0	0	Richard Pierce	Bot Mulcher
2	2	0	0	0:00:00	1	50	0:01:41	0	0	0	0	0	0	Eric Mallory	BotABing
2	2	0	0	0:00:00	1	50	0:00:58	1	6	2	2	2	0	Bruce Mobley	Botilla the Hun
2	4	2	50	0:01:42	1	25	0:02:00	1	2	1	0	1	4	Carl Moyer	Botknocker 19
2	1	0	0	0:00:00	0	0	0:00:00	1	14	5	4	5	0	Danis Carter	Bottom Dweller
1	2	0	0	0:00:00	0	0	0:00:00	2	20	6	9.5	4.5	2	Kevin Dahlquist	Botulism
1	4	2	50	0:00:45	0	0	0:00:00	2	23	8.5	8	6.5	6	Alan Martinson	Botulizer
2	7	0	0	0:00:00	1	14.28	0:01:40	6	29	9.17	10.5	9.33	10	Henry L. Hillman Jr.	Bouncing Betty
1	1	0	0	0:00:00	0	0	0:00:00	1	11	3	5	3	0	Nathan Roseborrough	Brain Mold
1	2	0	0	0:00:00	1	50	0:01:07	1	27	9	9	9	2	Francis Ryan	Brainsters of Triskellian, The
1	3	1	50	0:00:42	0	0	0:00:00	1	18	8	6	4	2	Jimmy Myers	Bucky The Beaverbot
3	5	1	20	0:03:00	1	20	0:01:26	3	23	9.33	6	7.67	4	Nola Garcia	Buddy Lee Don't Play in the Street
2	5	2	40	0:01:00	2	40	0:01:12	0	0	0	0	0	6	John Keyes	Bulldog
1	0	0	0	0:00:00	0	0	0:00:00	0	0	0	0	0	0		Bumble Bee
1	1	0	0	0:00:00	0	0	0:00:00	1	17	4	5	8	0	Tom Green	Bumper
2	2	0	0	0:00:00	2	100	0:01:32	0	0	0	0	0	0	Erik Hagman	BumperBawt
1	2	1	50	0:01:09	0	0	0:00:00	1	22	7	6	9	2	Brett Bellmore	Burning Metal
1	2	1	50	0:01:27	0	0	0:00:00	1	8	3	2	3	2	Tom Lombardi	Bust-O-Matic
2	4	0	0	0:00:00	1	25	0:03:25	1	31	12	12	7	4	David Norris	Caliban
1	1	0	0	0:00:00	0	0	0:00:00	1	18	6	6	6	0	James Miller	Cannibot
1	4	1	33.33	0:01:47	0	0	0:00:00	2	25	9.5	7	8.5	4	Doug Welch	Captain Insane-O
2	8	3	37.5	0:02:09	1	12.5	0:02:27	4	32	10.25	10	11.75	12	Bob Pitzer	Carnage Raptor
1	3	1	33.33	0:01:40	2	66.66	0:02:14	0	0	0	0	0	2	Ilya Polyakov	Carnivore
2	4	0	0	0:00:00	2	50	0:02:13	2	28.5	9.5	10.5	8.5	4	Thomas Beaver	Center Punch
1	1	0	0	0:00:00	0	0	0:00:00	1	19	6	5	8	0	William Arden	Centipede
1	3	0	0	0:00:00	1	33.33	0:01:50	1	26	8	9	9	4	Robert Davis	Cerberus
1	1	0	0	0:00:00	0	0	0:00:00	1	7	2	3	2	0	John Taylor	Chains Addiction
2	2	0	0	0:00:00	0	0	0:00:00	2	17.5	8.5	6	3	0	Greg Hjelstrom	Chameleon
1	3	1	33.33	0:01:20	1	33.33	0:01:01	1	24	10	7	7	4	Nathan Roseborrough	Cheap Shot
1	2	1t	50	0:01:02	0	0	0:00:00	1	10	5	2	3	2	Larry Beckman	Chewbo
1	1	0	0	0:00:00	1	100	0:01:23	0	0	0	0	0	0	Alexi Drosos	Chewy
1	1	0	0	0:00:00	1	20	0:00:58	3	19.33	6	7.33	6	4	Cassidy Wright	Chiabot
3	5	0	0	0:00:00	1	100	0:02:20	0	0	0	0	0	0	Cassidy Wright	ChiaBot
2	2	0	0	0:00:00	0	0	0:00:00	2	11.5	4	3	4.5	0	Tom Hall	Chimera

Robot Name	Team	Ranking	Weapon	W/L (W/L%)
Chipper Monkey	Chipper Monkey	#111 Middleweight	3000 RPM spinning lawn mower blade	0/1 (0.00%)
Chomp	Team LungFish	#112 Heavyweight	Chomping mouth; railroad pickax tail	0/1 (0.00%)
Chopper	Team Annihilator	#51 Middleweight	Horizontal rotating spring steel blade	2/2 (50.00%)
Chunky 357	The Chunky Monkeys	#178 Lightweight	High-speed rotating talons	0/1 (0.00%)
Circuit Breaker	Team Circuit Breaker	#75 Heavyweight	Spike	1/3 (25.00%)
Claymore	Claymore	#125 Lightweight	Centrifugal ring	0/3 (0.00%)
Cleprechaun	Cleprechaun	#110 Lightweight	Spinning blade	1/1 (50.00%)
Codebreaker	Diginati	#95 Super Heavyweight	Rotating steel drum	0/1 (0.00%)
Cold Chisel	C-M Engineering	#206 Lightweight	Invertable wedge	0/1 (0.00%)
Cold Steel	Tarantula	#179 Lightweight	Interchangeable vertical spinning disk or hammer	0/1 (0.00%)
Collateral Damage	Team Kinetic Energy	#87 Heavyweight	(top secret)	0/1 (0.00%)
Commando	Dirty Dog	#122 Middleweight	4500 RPM spinning disk; bulldozer	0/1 (0.00%)
Complete Control	Automatum	#5 Middleweight	Lifting/grabbing claw with giant opposable thumb	7/3 (70.00%)
Concretor	Deceptibots	#164 Lightweight	Opposable steel wedge reinforced with concrete	0/1 (0.00%)
Count Botula	Killer-bot	#137 Middleweight	Lifting clamp	0/1 (0.00%)
Crab Meat	Starboard!	#60 Heavyweight	Crushing claw	1/1 (50.00%)
Crash Test Dummy	Robot Workshop	#73 Heavyweight	Front/rear airbag	1/1 (50.00%)
Crashius Clay	Team Green Machine	#41 Lightweight	Interchangeable shafted weapons array	3/2 (60.00%)
Critical Condition	Team Ares	#127 Lightweight	1300 RPM spinning steel blade	0/2 (0.00%)
Crock	Crock	#107 Heavyweight		0/2 (0.00%)
Crouching Puppy, Hidden Kitten	Team CPHK	#96 Middleweight	Invertible dual wedge	1/1 (50.00%)
Crusher	Team Attitude	#15 Lightweight	Wedge	6/6 (50.00%)
CU-Denver Demagogue	Demagogue	#40 Middleweight	Two 10″ horizontal saw blades running at 5000 RPM	3/1 (75.00%)
CUAD The Annihilator	Team CUAD	#93 Middleweight	Fast-driven wedge	1/1 (50.00%)
CUAD The Crusher	Team CUAD	#29 Super Heavyweight	Vertical 360-degree clamping, lifting, ramming, spinning arm	2/2 (50.00%)
CUAD The Unmerciful	Team CUAD	#113 Heavyweight	Two lifting arms, invertible wedge design	0/1 (00.00%)
Cyclone	RKDx2	#78 Super Heavyweight	Rotating cyclonic blades	0/1 (0.00%)
The Cyclonic Plague	Team G.R.U.N.T.	#145 Middleweight		0/1 (0.00%)
Daisy	Team Daisy	#48 Lightweight	Spinning flower of doom	2/2 (50.00%)
Dark Matter	Cosmos	#152 Lightweight	Lifting arm; wedge	0/1 (0.00%)
Dark Steel	Team Knowledge	#180 Lightweight	Modular drill assembly; shovel force; wedge surfaces	0/1 (0.00%)
Darkness	Hardcore	#43 Middleweight	Rotating pickax	3/1 (75.00%)
Das Bot	Team Das Bot	#16 Lightweight	CO_2-powered lifting arm	5/3 (62.50%)
Dawn of Destruction	Klein	#25 Super Heavyweight	Lifting wedge ramp	3/1 (75.00%)
Deadblow	Team Deadblow	#2 Middleweight	Pneumatic hammer; flipper arm	7/6 (53.85%)
Death by Monkeys	Team Death by Monkeys	#8 Lightweight	Spikes	7/2 (77.78%)
Death Ray	Team Stingray	#142 Lightweight	Flipper arm	0/1 (0.00%)
Death-Maul	Team Patriot	#114 Heavyweight	2 masses spinning horizontally at the ends of a 4-5 ft. rotating arm	0/1 (0.00%)
Deathstar	Basher of Bots	#41 Heavyweight		2/1 (66.67%)
Deathtrap	Team Vicious	#35 Lightweight		4/2 (66.67%)

TT	TM	KOs	KO%	AKOT	KOA	KOA%	AKOAT	TJD	JPA	AA	DA	SA	Points	Builder	Robot Name
1	1	0	0	0:00:00	0	0	0:00:00	1	22	7	6	9	0	John Carioti	Chipper Monkey
1	1	0	0	0:00:00	1	100	0:01:45	0	0	0	0	0	0	Kevin Lung	Chomp
2	4	1	25	0:00:49	0	0	0:00:00	3	28.33	8.33	10	10	4	Adam Baxter	Chopper
1	1	0	0	0:00:00	1	100	0:01:48	0	0	0	0	0	0		Chunky 357
3	4	0	0	0:00:00	3	75	0:01:46	1	36	13	12	11	2	Josh O'Briant	Circuit Breaker
2	2	0	50	0:01:41	0	0	0:00:00	1	21	6	6	9	2	John Duncan	Claymore
1	2	0	0	0:00:00	0	0	0:00:00	1	18	7	5	6	0	Michael Clepper	Cleprechaun
1	1	0	0	0:00:00	1	100	0:01:04	0	0	0	0	0	0	Robert Sica	Codebreaker
1	0	0	0	0:00:00	0	0	0:00:00	0	0	0	0	0	0	Bill Olson	Cold Chisel
1	1	0	0	0:00:00	1	100	0:01:23	0	0	0	0	0	0	Frank Harris	Cold Steel
1	1	0	0	0:00:00	0	0	0:00:00	1	19	6	6	7	0	Mike Pelstring	Collateral Damage
1	1	0	0	0:00:00	0	0	0:00:00	1	17	5	6	6	0	Greg Watts	Commando
3	10	2	20	0:01:56	3	30	0:02:22	5	33.6	10.8	10.6	12.2	15	Derek Young	Complete Control
1	1	0	0	0:00:00	0	0	0:00:00	0	0	0	0	0	0	Joe Sena	Concretor
1	1	0	0	0:00:00	0	0	0:00:00	1	9	5	2	2	0	Alan Martinson	Count Botula
1	2	1	50	0:01:11	1	50	0:01:43	0	0	0	0	0	2	Don Harper	Crab Meat
1	2	0	0	0:00:00	0	0	0:00:00	2	19	6	6.5	6.5	2	Robert Doerr	Crash Test Dummy
2	5	0	0	0:00:00	0	0	0:00:00	5	25.2	8	9.2	8	6	Steven DeZwart	Crashius Clay
2	2	0	0	0:00:00	0	0	0:00:00	1	20	7	7	6	0	Robert Moore	Critical Condition
1	2	0	0	0:00:00	2	100	0:00:42	0	0	0	0	0	0	Gary Cline	Crock
1	2	0	0	0:00:00	0	0	0:00:00	2	20.5	7.5	6.5	6.5	2	Shaun Fetherston	Crouching Puppy, Hidden Kitten
5	12	4	33.33	0:01:54	3	25	0:01:53	1	15	5	5	5	10	Thomas Petruccelli	Crusher
1	4	0	0	0:00:00	0	0	0:00:00	4	26.5	9.25	7.75	9.5	6	Michael Gallaway	CU-Denver Demagogue
1	2	0	0	0:00:00	0	0	0:00:00	2	23	7	6.5	9.5	2	Matt Ulrey	CUAD The Annihilator
2	4	1	25	0:00:45	0	0	0:00:00	3	24.67	8	8	8.67	4	Matt Ulrey	CUAD The Crusher
1	1	0	0	0:00:00	1	100	0:01:18	0	0	0	0	0	0	Matt Ulrey	CUAD The Unmerciful
1	1	0	0	0:00:00	0	0	0:00:00	1	15	5	5	5	0	Kevin Knoedler	Cyclone
1	1	0	0	0:00:00	0	0	0:00:00	1	6	1	2	3	0	Jonathan Grimm	Cyclonic Plague, The
2	4	2	50	0:01:20	0	0	0:00:00	2	16.5	6	4	6.5	4	Derek Zahn	Daisy
1	1	0	0	0:00:00	0	0	0:00:00	1	6	2	3	1	0	Peter Covert	Dark Matter
1	1	0	0	0:00:00	1	100	0:01:15	0	0	0	0	0	0	Russ Barrow	Dark Steel
1	4	0	0	0:00:00	1	25	0:02:37	2	33.5	11	11	11.5	6	Ray Billings	Darkness
3	8	3	37.5	0:01:10	1	12.5	0:00:57	4	27	9.25	8.5	9.25	10	Paul Mathus	Das Bot
1	4	0	0	0:00:00	1	25	0:02:01	3	26.67	8.33	9.33	9	6	Jim Klein	Dawn of Destruction
5	13	3	23.07	0:01:42	3	23.07	0:01:14	3	22.67	8.33	7	7.33	18	Grant Imahara	Deadblow
2	9	4	44.44	0:02:05	2	22.22	0:02:21	2	27	9.5	8	9.5	14	Robert Farrow	Death By Monkeys
1	1	0	0	0:00:00	0	0	0:00:00	1	13	6	6	1	0	Simon Arthur	Death Ray
1	1	0	0	0:00:00	1	100	0:00:44	0	0	0	0	0	0	Eric Twigg	Death-Maul
1	3	1	33.33	0:01:36	0	0	0:00:00	1	8	5	2	1	4	Wray Russ	Deathstar
1	6	2	33.33	0:02:12	0	0	0:00:00	0	0	0	0	0	6	Mike Regan	Deathtrap

Robot Name	Team	Ranking	Weapon	W/L (W/L%)
Deathwishbone	Tesla's Stooges	#115 Heavyweight	Spinning plates	0/1 (0.00%)
Deb Bot	Fast Electric Robots	#29 Middleweight	Lifting arm; ramming power	4/2 (66.67%)
Decimator	Decimator	#91 Heavyweight	CO_2-powered flipper arm	0/2 (0.00%)
Defiant	Defiant	#13 Lightweight		5/1 (83.33%)
Delta II	Macomb Mathematic Science Tech Center	#181 Lightweight	Reciprocating ram	0/1 (0.00%)
Deus Ex Machina	Team Scrounge Lizards	#106 Middleweight	Spring-powered flipping arm	1/1 (50.00%)
Deviation Response	Team Polafia	#138 Lightweight	Triangular ramming wedge; rear spike plate	0/1 (0.00%)
DiamondBack	I'm That Good	#159 Lightweight	Mini-wedge; spikes	0/2 (0.00%)
Die Fledermaus	Emergency Third Rail Power Trip	#99 Middleweight	Ramming power	1/1 (50.00%)
Diesector	Mutant Robots	#2 Super Heavyweight	Twin chisels; hammers; the jaws of death	11/3 (78.57%)
Disco Cow 3000	Little People	#116 Heavyweight	Super robot strength	0/1 (0.00%)
Disposable Hero	Lungfish Technologies	#182 Lightweight		0/1 (0.00%)
Dispose-All	Team LOGICOM	#139 Middleweight	Leverage; kinetic energy	0/1 (0.00%)
DooAll	Team DooAll	#13 Super Heavyweight	CO_2-powered ram	4/4 (50.00%)
Doom of Babylon	Team Revelation	#49 Super Heavyweight	Kinetic energy	1/1 (50.00%)
The Doomsday Machine	The Doom Crew	#149 Lightweight		0/1 (0.00%)
Doorstop	Team Malfeasance	#149 Middleweight	Wedge	0/1 (0.00%)
Double Agent	Diginati	#24 Middleweight	Double-wedge design; 6 titanium spikes	4/2 (66.67%)
Double Secret Probation	Hamncheez	#147 Lightweight	Wedge	0/1 (0.00%)
Double Trouble	Killer BotZ	#166 Middleweight	Front and back flippers; bimodal wedge	0/1 (0.00%)
Doublecross	Team Doublecross	#183 Lightweight	2 spinning blades	0/1 (0.00%)
Dr. Inferno Jr.	The Infernolab	#5 Lightweight		7/3 (70.00%)
Dracolich	Team Crash	#113 Lightweight	61-cc. chainsaw; mini-wedge skirts	1/1 (50.00%)
Dreadbot	Silicon Valley Destruction Company	#35 Super Heavyweight	Inertia	2/3 (40.00%)
Dreadnought	Team Coolrobots	#23 Super Heavyweight		3/2 (60.00%)
Dredge	Lancers	#138 Heavyweight	Lances and steel mace	0/1 (0.00%)
Drill-O-Dillo	Born2bot	#96 Super Heavyweight	Rotating 6-in. tri-cone drill bit; wedge-shaped tail	0/1 (0.00%)
Dude of Destruction	Lightning	#122 Lightweight	3 spinning 7-in. saw blades; changeable horizontal spikes, wedges, or rams	0/1 (0.00%)
The Dude	Explosivo	#58 Super Heavyweight	6 impellers spinning at up to 4000 RPM	1/1 (50.00%)
El Cucuy	NEXUS6	#167 Middleweight	Hammer pick	0/1 (0.00%)
El Diablo	Team Diablo	#10 Middleweight	Trench saw with a large toothed steel drum	6/2 (75.00%)
El Diablo Grande	Team Diablo	#67 Heavyweight	Spinning toothed barrel with gut rippers	1/2 (33.33%)
Electric Lunch	Team K.I.S.S.	#14 Super Heavyweight	Lifting arm; ramming power	4/3 (57.14%)
The Emasculator	Team Wetware	#198 Middleweight	6-ft. helicopter blade	0/2 (0.00%)
Endotherm	Team Saber	#50 Lightweight	Momentum	3/3 (50.00%)
Enforcer	Half Fast Astronaut	#103 Lightweight	Lifter and saw arm	1/1 (50.00%)
EnumScratch	BTB	#140 Middleweight	Spinning Thing of Death	0/1 (0.00%)
Eradicator	Absolute	#55 Super Heavyweight	Titanium spike; lifting/flipping arm	1/2 (33.33%)

TT	TM	KOs	KO%	AKOT	KOA	KOA%	AKOAT	TJD	JPA	AA	DA	SA	Points	Builder	Robot Name
1	1	0	0	0:00:00	1	100	0:01:28	0	0	0	0	0	0	Chad Castillo	Deathwishbone
2	6	1	20	0:01:04	1	20	0:01:26	3	26.33	9.67	7.67	9	6	Jeff Vasquez	Deb Bot
2	2	0	0	0:00:00	0	0	0:00:00	2	8.5	5.5	1.5	1.5	0	Kyle Vogt	Decimator
1	6	3	50	0:02:56	0	0	0:00:00	0	0	0	0	0	11	James Underwood	Defiant
1	1	0	0	0:00:00	1	100	0:02:44	0	0	0	0	0	0	Mark Supal	Delta II
1	2	0	0	0:00:00	1	50	0:01:20	0	0	0	0	0	2	Gary Slack	Deus Ex Machina
1	1	0	0	0:00:00	0	0	0:00:00	1	15	4	6	5	0	Dave Morris	Deviation Response
2	1	0	0	0:00:00	1	100	0:01:10	0	0	0	0	0	0	Peter McNeil	DiamondBack
1	2	0	0	0:00:00	0	0	0:00:00	2	19	6	7.5	5.5	2	Blake Mitchell	Die Fledermaus
4	14	6	42.85	0:01:50	1	7.14	0:01:36	7	30.57	10.57	9.29	10.71	24	Donald Hutson	Diesector
1	1	0	0	0:00:00	1	100	0:01:24	0	0	0	0	0	0	Patrick Dwight	Disco Cow 3000
1	1	0	0	0:00:00	1	100	0:00:50	0	0	0	0	0	0	Phil Putman	Disposable Hero
1	1	0	0	0:00:00	0	0	0:00:00	1	8	3	3	2	0	Brian Nave	Dispose-All
4	7	2	28.57	0:02:18	0	0	0:00:00	5	15.4	6	5.6	3.8	9	Scott LaValley	DooAll
1	2	1	50	0:01:17	0	0	0:00:00	1	21	7	7	7	2	Todd Mendenhall	Doom of Babylon
1	1	0	0	0:00:00	0	0	0:00:00	1	9	3	1	5	0	Terry Wynn	Doomsday Machine, The
1	1	0	0	0:00:00	0	0	0:00:00	1	2	1	1	0	0	David Pabian	Doorstop
2	6	3	50	0:01:39	0	0	0:00:00	3	13.67	4.67	4.33	4.67	8	Robert Sica	Double Agent
1	1	0	0	0:00:00	0	0	0:00:00	1	11	1	4	6	0	Bryson DeJong	Double Secret Probation
1	1	0	0	0:00:00	1	100	0:01:20	0	0	0	0	0	0	Edward Haas	Double Trouble
1	1	0	0	0:00:00	1	100	0:01:04	0	0	0	0	0	0	Andrew Schwirian	Doublecross
4	10	2	20	0:01:36	1	10	0:01:38	6	27.17	9.5	7.83	9.83	18	Jason Bardis	Dr. Inferno Jr.
1	2	0	0	0:00:00	1	50	0:00:45	0	0	0	0	0	2	William Bottenberg	Dracolich
3	5	1	20	0:01:30	1	20	0:00:48	3	22	8.67	6.67	6.67	4	David Hall	Dreadbot
2	5	1	20	0:01:48	1	20	0:01:28	3	27	8.33	8.33	10.33	6	Christian Carlberg	Dreadnought
1	0	0	0	0:00:00	0	0	0:00:00	0	0	0	0	0	0	Gene Burnett	Dredge
1	1	0	0	0:00:00	1	100	0:01:32	0	0	0	0	0	0	Andrew Gardner	Drill-O-Dillo
1	1	0	0	0:00:00	0	0	0:00:00	1	22	8	8	6	0	Steven Karp	Dude of Destruction
1	1	0	0	0:00:00	0	0	0:00:00	1	30	9	8	13	2	Patrick Urschel	Dude, The
1	1	0	0	0:00:00	1	100	0:01:41	0	0	0	0	0	0	Gerardo Serafin	El Cucuy
2	8	4	50	0:02:42	1	12.5	0:00:45	2	25.5	8	7.5	10	13	Zach Bieber	El Diablo
2	3	1	33.33	0:02:00	0	0	0:00:00	2	17	6	6.5	4.5	2	Zach Bieber	El Diablo Grande
3	6	2	33.33	0:01:27	0	0	0:00:00	4	23.75	8.5	7.5	7.75	8	Steven Nelson	Electric Lunch
2	0	0	0	0:00:00	0	0	0:00:00	0	0	0	0	0	0	Ray Scully	Emasculator, The
2	6	2	33.33	0:02:31	1	16.66	0:04:30	2	25	10	7.5	7.5	4	John Patrick	Endotherm
1	2	0	0	0:00:00	0	0	0:00:00	2	18	4.5	5.5	8	2	Al Kindle	Enforcer
1	1	0	0	0:00:00	0	0	0:00:00	1	8	3	2	3	0	Brian Patrick	EnumScratch
2	3	0	0	0:00:00	1	33.33	0:02:05	2	25.5	8	9.5	8	2	Ken Gentry	Eradicator

Robot Name	Team	Ranking	Weapon	W/L (W/L%)
Eraser	Pencil Neck Geeks	#15 Middleweight	Spike	5/2 (71.43%)
Eric the Red	Team Viking USA	#129 Lightweight	Twin double-headed ax blades	0/2 (0.00%)
Evil Con Carne	Team DeSade	#59 Middleweight	Overhead hammer	2/1 (66.67%)
Evil Fish Tank	Team Delta	#65 Lightweight	Actuated titanium ramp	2/1 (66.67%)
Excalibot	Merlin	#97 Super Heavyweight	Pneumatic ram; horsepower and torque	0/1 (0.00%)
Executioner	Team SMC	#22 Lightweight		6/2 (75.00%)
Exodus 2001	Exodus	#42 Middleweight	Rotary impact; pushing power	3/2 (60.00%)
Exterminator	The Exterminator	#139 Heavyweight		0/2 (0.00%)
Eye of the Storm	Team Storm	#98 Middleweight	Spiked mace	1/1 (50.00%)
F5	Unidentified Flying Object	#16 Middleweight	1.5-lb. blades, spinning at 2000 RPM	5/1 (83.33%)
Fang	Team Bulldog	#56 Lightweight	Steel pincer	2/2 (50.00%)
Fantom Thrust	Fantom Motion Inc.	#43 Super Heavyweight	Hydraulic ram with tool steel spike	2/1 (66.67%)
Fast Forward	Shaped Reality	#69 Middleweight	High-speed spinning weapon	2/2 (50.00%)
Firestorm		#107 Lightweight	Stationary spears	1/1 (50.00%)
Flailbot	Downhill	#47 Super Heavyweight	Spinning flails and rear battering ram	1/1 (50.00%)
FlapJack	JohnsenVille Bots	#20 Middleweight	Lifting arm flipper powered by compressed air. Low pressure pneu at 250psi, but w/high flow valve; 3" Bore cylinder.	0/1 (0.00%)
FlatBot	Team Theta	#73 Lightweight	Saw blade	1/1 (50.00%)
Flip You Off	Team Squat Bot	#168 Middleweight	Pneumatic flipper	0/1 (0.00%)
Fork-N-Stein	American Performance Robotics	#77 Heavyweight	Lifting forks	1/2 (33.33%)
frenZy	Team Minus Zero	#10 Heavyweight	Ax	7/7 (50.00%)
Froghemoth	Ontave	#184 Lightweight	Rotating Shell of Ambivalence	0/1 (0.00%)
FrostBite	Team Toad	#30 Heavyweight	Snowplow of Death	3/3 (50.00%)
Frozen Toast	Team Evil Duck	#185 Lightweight	Saw	0/1 (0.00%)
Fusion	Team Saber	#154 Middleweight	Pneumatic punch/spear	0/2 (0.00%)
Gamma Raptor	Team Raptor	#3 Lightweight	Lifter arm	11/4 (73.33%)
Gammacide	Gammatronic Robot Brigade	#21 Super Heavyweight	Scoop	3/1 (75.00%)
Gammatron	Gammatronic Robot Brigade	#71 Heavyweight	Hoop of Hopeless Doom	1/3 (25.00%)
Gang Green Weasel	Team Underground	#65 Heavyweight	Speed; acceleration; weight (200 lbs.)	1/1 (50.00%)
Garm	Kampfgruppe5	#56 Middleweight	Vertical spinner	2/1 (66.67%)
Gary Gizzmo	Kampfgruppe5	#94 Middleweight	Wedge and twin swinging hammers	1/1 (50.00%)
General Gau	Team Goosebearys	#45 Lightweight	Strong drive system; ultra-grip tires	3/2 (60.00%)
Gerbilov	Stendeck	#165 Lightweight	Spinning bowl	0/1 (0.00%)
Ginsu	La Ma Motors	#62 Super Heavyweight		1/3 (25.00%)
Give Me Some Sugar Baby	Team Sugar	#110 Heavyweight	Pneumatic double fork	0/1 (0.00%)
GoatHammer	The Goat Brothers	#119 Middleweight	Pry bar spears; rigid-mounted saw blades on wedge	0/1 (0.00%)
Godtilla	Lunchbreak	#186 Lightweight	Spinner	0/1 (0.00%)
GoldDigger	Team Boomer	#20 Heavyweight	Pickax	4/7 (36.36%)
GoPostal	Post Masters	#70 Super Heavyweight	Rotating steel hammers	0/1 (0.00%)
Gorange 3	Team Deranged	#69 Lightweight		2/2 (50.00%)

TT	TM	KOs	KO%	AKOT	KOA	KOA%	AKOAT	TJD	JPA	AA	DA	SA	Points	Builder	Robot Name
2	7	4	57.14	0:01:15	1	14.28	0:01:35	2	23	7	8	8	10	Kenneth Fuiks	Eraser
2	2	0	0	0:00:00	0	0	0:00:00	2	9.5	4.5	2	3	0	Eric Dalnes	Eric The Red
1	3	1	33.33	0:01:58	0	0	0:00:00	2	20.5	7	6.5	7	4	Terry Talton	Evil Con Carne
1	3	0	0	0:00:00	0	0	0:00:00	2	19.5	5.5	8	6	4	Dan Danknick	Evil Fish Tank
1	1	0	0	0:00:00	1	100	0:01:11	0	0	0	0	0	0	Wayne Sommars	Excalibot
1	8	3	37.5	0:02:15	2	25	0:02:33	0	0	0	0	0	8	Andrew Duerner	Executioner
2	5	0	0	0:00:00	1	20	0:04:11	3	29	9	11.33	8.67	6	Patrick Verner	Exodus 2001
1	0	0	0	0:00:00	0	0	0:00:00	0	0	0	0	0	0	Geoffrey Chandler	Exterminator
1	2	0	0	0:00:00	0	0	0:00:00	2	20	6	5.5	8.5	2	Jonathan Bedford	Eye of the Storm
1	6	4	66.66	0:02:08	1	16.66	0:01:16	1	28	8	12	8	10	Dan Bovinich	F5
2	4	1	25	0:01:23	1	25	0:00:57	2	28.5	9	10.5	9	4	Brian Foote	Fang
1	3	0	0	0:00:00	1	50	0:01:22	1	27	8	9	10	2	Michael J. Kawecki	Fantom Thrust
2	4	0	0	0:00:00	1	33.33	0:01:14	2	27	9.5	9	8.5	2	Phil Houdek	Fast Forward
1	2	0	0	0:00:00	0	0	0:00:00	1	25	7	8	10	2	Tom Burbeck	Firestorm
1	2	1	50	0:00:52	0	0	0:00:00	1	15	4	5	6	2	John Phillips	Flailbot
1	0	0	0	0:00:00	0	0	0:00:00	0	0	0	0	0	0	Mark Johnsen	FlapJack
1	2	1	50	0:00:57	1	50	0:01:08	0	0	0	0	0	2	Randal Boyd	FlatBot
1	1	0	0	0:00:00	1	100	0:00:45	0	0	0	0	0	0	Ken Stevens	Flip You Off
2	3	0	0	0:00:00	1	33.33	0:02:08	1	16	6	5	5	2	Jim Snook	Fork-N-Stein
6	14	4	28.57	0:01:22	3	21.42	0:02:26	4	23.5	6.25	8.25	9	13	Patrick Campbell	frenZy
1	1	0	0	0:00:00	1	100	0:01:08	0	0	0	0	0	0	James Thomas	Froghemoth
3	6	1	16.66	0:02:20	1	16.66	0:01:36	4	23.25	8	7.25	8	6	Michael Mauldin	FrostBite
1	1	0	0	0:00:00	1	100	0:00:49	0	0	0	0	0	0		Frozen Toast
2	2	0	0	0:00:00	2	100	0:02:16	0	0	0	0	0	0	John Patrick	Fusion
4	15	5	33.33	0:01:24	3	20	0:01:52	7	31.43	10.71	11	9.71	24	Chuck Pitzer	Gamma Raptor
1	4	2	50	0:01:11	1	25	0:01:58	1	29	10	11	8	6	Mike Okerman	Gammacide
3	4	0	0	0:00:00	2	50	0:02:00	2	20	0	12.5	7.5	2	Mike Okerman	Gammatron
1	2	1	50	0:01:28	1	50	0:01:38	0	0	0	0	0	2	Michael Springer	Gang Green Weasel
1	3	1	33.33	0:01:25	1	33.33	0:01:02	1	34	12	12	10	4	Mike Flanagan	Garm
1	2	0	0	0:00:00	0	0	0:00:00	2	22	7	6.5	8.5	2	Mike Flanagan	Gary Gizzmo
2	5	0	0	0:00:00	2	40	0:01:38	2	26	9	8	9	6	William Garcia	General Gau
1	1	0	0	0:00:00	0	0	0:00:00	0	0	0	0	0	0	John Loo	Gerbilov
3	4	0	0	0:00:00	1	25	0:02:45	1	10	10	0	0	2	Trey Roski	Ginsu
1	1	0	0	0:00:00	0	0	0:00:00	0	0	0	0	0	0	Clyde Ruckle	Give Me Some Sugar Baby
1	1	0	0	0:00:00	0	0	0:00:00	1	18	5	7	6	0	Timothy Ray	GoatHammer
1	1	0	0	0:00:00	1	100	0:01:11	0	0	0	0	0	0	Anthony Speros	Godtilla
5	8	1	12.5	0:01:14	2	25	0:01:30	5	23.8	6.8	8.4	8.6	8	Randy Eubanks	GoldDigger
1	1	0	0	0:00:00	0	0	0:00:00	1	20	7	7	6	0	John Taylor	GoPostal
1	4	0	0	0:00:00	2	50	0:02:18	0	0	0	0	0	3	Grayson DuRaine	Gorange 3

Robot Name	Team	Ranking	Weapon	W/L (W/L%)
G.O.R.T.	Lone Star Panthers	#107 Super Heavyweight	Powered spike	0/1 (0.00%)
GrayMatter	Team GrayMatter	#50 Super Heavyweight	16-in. stainless steel spike	1/3 (25.00%)
Green Dragon	Team Dragon Combat Robots	#154 Lightweight	Spinning saw blade; double wedge design	0/2 (0.00%)
Greenspan	Team Brown	#25 Heavyweight	24" Hammermill	3/2 (60.00%)
Gremlin II	Team Duct Tape	#187 Lightweight		0/1 (0.00%)
Grunion	Terminator	#83 Lightweight		1/1 (50.00%)
Gungnire	Pilgrim Nuclear Technology Alliance	#100 Middleweight	Pusher/basher	1/1 (50.00%)
Gyrax	Razorback	#67 Super Heavyweight	Rotating hammer and slicers	0/2 (0.00%)
Half Ass N Random	Measure Once, Cut Twice	#72 Lightweight	Blunt Force Trauma	1/1 (50.00%)
Half Gassed	S.L.A.M	#60 Super Heavyweight	Push-type wedge	1/2 (33.33%)
Halo	Team Second Breakdown	#169 Middleweight	Spinning halo flails	0/1 (0.00%)
HammerHead	Team Brute Force	#29 Lightweight	Inertia	2/4 (33.33%)
Hammerlock	Bots of Wrath	#108 Super Heavyweight	32-inch steel spinner driven by 5.5hp Honda	0/1 (0.00%)
Hammertime	Team Hammertime	#31 Super Heavyweight	14-in. pneumatic steel hammer spike	2/2 (50.00%)
Hamunaptra	Team Codisys	#98 Super Heavyweight	12-lb. spiked spinning balls	0/1 (0.00%)
Harvey Wall Banger	Harvey Wall Banger	#116 Lightweight		1/2 (33.33%)
Hazard	Team Hazard	#1 Middleweight	Spinning Blade of Death	14/0 (100%)
Hazardous Waste of Time	Loyal Robot Army	#117 Heavyweight	Horizontal/vertical spinner	0/1 (0.00%)
Headtrimmer	Springcrest	#188 Lightweight	Reciprocating saws on ramps and sides	0/1 (0.00%)
The Heart of Darkness	Apocalypse Now	#63 Heavyweight		1/1 (50.00%)
Heavy Metal Noise	Big Bang Robotics	#13 Middleweight	Twin 4000 RPM vertical spinners	6/1 (85.71%)
Herr Gepoünden	Team Zwölfpack	#17 Lightweight	Unpleasant demeanor; wildly flailing heavy thingie	5/2 (71.43%)
Hexadecimator	Team WhoopAss	#7 Heavyweight	Lifting arm	7/2 (77.78%)
Hexy Jr.	Team WhoopAss Jr.	#14 Lightweight	Flipping arm	5/1 (83.33%)
High Impact	The Armada	#41 Super Heavyweight	Gas-powered rotating shell with welded blades	2/2 (50.00%)
Hipopononomus	Team Batteraizees	#170 Middleweight	16.5-in. toothed aluminum flywheel	0/1 (0.00%)
Hobgoblin LTD	Team Warlock	#133 Lightweight	Saw-toothed hammertail	0/1 (0.00%)
Hoobot	Team Ballyhoo	#42 Super Heavyweight	Flipping front end	2/2 (50.00%)
Horrifica	Team Beta	#54 Middleweight	Speed	2/1 (66.67%)
Hot Air	Lungfish Technologies	#114 Lightweight		1/2 (33.33%)
Huggy Bear	Team Huggy Bear	#17 Middleweight	Pneumatic squeezing mechanism	5/2 (71.43%)
Hurricane	TU Women's Robotics	#89 Super Heavyweight	Metal spikes attached to a spinning body	0/1 (0.00%)
I-Beam	Team Girder	#28 Heavyweight	Spinning or ramming	3/2 (60.00%)
Ice Cube	Team Crystal	#38 Middleweight	Reinforced steel snowplow blade	3/2 (60.00%)
IceBerg	Team Toad	#16 Super Heavyweight	Pneumatic, knife-edged steel snowplow blade	4/2 (66.67%)
Icky Thump	Icky Thump	#70 Heavyweight	5-in. spinning barrel with 2-in. spikes	1/1 (50.00%)
Ill Intent	Ill Intent	#202 Middleweight		0/2 (0.00%)
The Immortal Chaos	The Immortals	#93 Heavyweight	Large spinning wheel with teeth	0/1 (0.00%)
Incisor	Pain USA	#39 Heavyweight	Twin lift arms within a forward wedge and spiked hammer	2/2 (50.00%)
Incoming	Anger Management Inc.	#16 Heavyweight	3 serrated flails hinged on a 100-lb. 500 RPM spinning steel disk	4/2 (66.67%)

TT	TM	KOs	KO%	AKOT	KOA	KOA%	AKOAT	TJD	JPA	AA	DA	SA	Points	Builder	Robot Name
1	0	0	0	0:00:00	0	0	0:00:00	0	0	0	0	0	0	Kelly Osborn	G.O.R.T.
3	4	1	25	0:01:32	1	25	0:01:50	2	12	2.5	1.5	8	2	Israel Mathewson	GrayMatter
2	2	0	0	0:00:00	2	100	0:01:50	0	0	0	0	0	0	John Pagano	Green Dragon
2	5	2	40	0:01:22	2	40	0:01:15	1	27	9	9	9	6	Steven Brown	Greenspan
1	1	0	0	0:00:00	1	100	0:02:20	0	0	0	0	0	0	Jerome Miles	Gremlin II
1	2	1	50	0:01:32	1	50	0:01:10	0	0	0	0	0	2	K. Brian O'Neal	Grunion
1	2	0	0	0:00:00	1	50	0:01:58	1	23	7	7	9	2	Michael Bastoni	Gungnire
2	2	0	0	0:00:00	0	0	0:00:00	2	16	5	5.5	5.5	0	David Peavy	Gyrax
1	2	1	50	0:00:48	1	50	0:01:39	0	0	0	0	0	2	David Robison	Half Ass N Random
2	3	0	0	0:00:00	2	66.66	0:01:11	1	24	8	8	8	2	Lowell Nelson	Half Gassed
1	1	0	0	0:00:00	1	100	0:01:09	0	0	0	0	0	0	Brian Scearce	Halo
3	6	1	16.66	0:02:31	2	33.33	0:01:50	2	22.5	7.5	7.5	7.5	8	Charles Steinkuehler	HammerHead
1	0	0	0	0:00:00	0	0	0:00:00	0	0	0	0	0	0	Tom Wickey	Hammerlock
2	4	1	25	0:01:07	1	25	0:01:07	2	23	6	8	9	4	Jerry Clarkin	Hammertime
1	1	0	0	0:00:00	1	100	0:00:51	0	0	0	0	0	0	Nathan Poort	Hamunaptra
1	3	0	0	0:00:00	2	66.66	0:03:23	0	0	0	0	0	1	Robert Anderson	Harvey Wall Banger
3	14	10	71.42	0:01:13	0	0	0:00:00	4	35.25	10.75	13.25	11.25	36	Tony Buchignani	Hazard
1	1	0	0	0:00:00	1	100	0:00:43	0	0	0	0	0	0	Steven Lawver	Hazardous Waste Of Time
1	1	0	0	0:00:00	1	100	0:01:41	0	0	0	0	0	0	Paul Bell	Headtrimmer
1	2	1	50	0:01:25	1	50	0:01:04	0	0	0	0	0	2	Steven Lawver	Heart of Darkness, The
1	7	4	57.14	0:01:55	1	14.28	0:00:37	1	36	14	14	8	12	Jay Johnson	Heavy Metal Noise
2	7	2	28.57	0:02:24	0	0	0:00:00	5	23.6	8.4	8.6	6.6	10	David Otto	Herr Gepoünden
2	9	5	55.55	0:01:42	2	22.22	0:01:44	2	27.5	8.5	9.5	9.5	16	Tim Paterson	Hexadecimator
1	6	5	83.33	0:01:13	1	16.66	0:02:58	0	0	0	0	0	10	Scott Ferguson	Hexy Jr.
2	4	0	0	0:00:00	1	25	0:02:38	3	25.33	9	7.67	8.67	4	Victor Garaycochea	High Impact
1	1	0	0	0:00:00	1	100	0:01:21	0	0	0	0	0	0	Justin Chou	Hipopononomus
1	1	0	0	0:00:00	0	0	0:00:00	1	17	6	6	5	0	Chad Campbell	Hobgoblin LTD
2	4	0	0	0:00:00	1	25	0:00:41	3	20.33	6.67	7	6.67	4	Rocky Mohacsi	Hoobot
1	3	1	33.33	0:01:20	1	33.33	0:01:04	1	27	9	9	9	4	Brian Patrick	Horrifica
1	2	0	0	0:00:00	1	50	0:00:53	0	0	0	0	0	2	Phil Putman	Hot Air
2	7	2	28.57	0:01:13	1	14.28	0:01:42	4	30.5	9.75	9.75	11	10	Dave Schultz	Huggy Bear
1	1	0	0	0:00:00	0	0	0:00:00	1	0	0	0	0	0	Doug Jussaume	Hurricane
2	5	1	20	0:01:45	2	40	0:01:07	2	30.5	9	9	12.5	6	Dean DuBois	I-Beam
2	5	1	20	0:01:15	0	0	0:00:00	4	25.5	9	8	8.5	6	Debbie Mauldin	Ice Cube
2	6	1	16.66	0:00:36	0	0	0:00:00	5	21	6.8	6.6	7.6	8	Michael Mauldin	IceBerg
1	2	0	0	0:00:00	0	0	0:00:00	2	22	8	6.5	7.5	2	Mike Olsen	Icky Thump
1	0	0	0	0:00:00	0	0	0:00:00	0	0	0	0	0	0	Morgan Dunbar	Ill Intent
1	1	0	0	0:00:00	0	0	0:00:00	1	16	7	6	3	0	Charles Constantine	Immortal Chaos, The
2	4	1	25	0:01:15	1	25	0:01:57	1	16	5	4	7	4	Kevin McEnery	Incisor
2	6	2	33.33	0:01:32	0	0	0:00:00	4	15.5	5.5	5	5	8	Brian Moore	Incoming

Robot Name	Team	Ranking	Weapon	W/L (W/L%)
Instigator	Team Byte Me	#104 Middleweight	Kinetic energy disk with teeth	1/4 (20.00%)
Interceptor	Absolute Zero	#118 Heavyweight	Lifting arm	0/1 (0.00%)
Internal Audit	Team Travesty	#171 Middleweight	Horizontal spinning blade	0/1 (0.00%)
InTriVerter	Team Think Tank	#89 Heavyweight	10-lb. hammer	0/1 (0.00%)
Iron Eagle	Nishimura Robot Club	#119 Heavyweight	Beak of the Eagle	0/1 (0.00%)
Iron Soldier	Team ACMX	#87 Middleweight	Wedge with actuated lifting rails	1/2 (33.33%)
Isosceles 1	Team Trigon	#99 Super Heavyweight	8-in. overhead steel pick	0/1 (0.00%)
I-Will-Smack-U	Shuddup	#47 Lightweight	Pneumatic lifting arm	2/1 (66.67%)
Jabberwock	Robot Warriors	#22 Heavyweight	Invertible, double-wedge; pneumatic ram	3/1 (75.00%)
Jack the Flipper	Methodical Robots	#105 Lightweight	Spring-powered titanium flipper arm	1/1 (50.00%)
Jack the Gripper	Get a Grip	#100 Super Heavyweight	Flipping/crushing/lifting arm	0/1 (0.00%)
The Jack Tripper	Measure Once, Cut Twice	#120 Heavyweight		0/1 (0.00%)
Janus	Team Janus	#54 Super Heavyweight	Universal positioned wedge; full rear battering ram surface	1/1 (50.00%)
Jaws of Death	Operation Boilermaker	#34 Super Heavyweight	Large hydraulic jaw	2/2 (50.00%)
Jersey Devil	Jersey Devil	#123 Middleweight	Spinning wedge	0/1 (0.00%)
Jo Mama	Team Flaming Monkeys	#43 Lightweight	Lifting arm	3/2 (60.00%)
Jolly Roger	Team Tenacity	#121 Heavyweight	CO_2-powered spike	0/1 (0.00%)
The Judge	Mechanicus	#9 Super Heavyweight		4/2 (66.67%)
JuggerBot	JuggerBot	#46 Super Heavyweight		1/1 (50.00%)
Juggernaut	Mechanicus	#51 Heavyweight		2/2 (50.00%)
Junior	Team Nightmare	#45 Middleweight		3/3 (50.00%)
The Junkyard Cog	The Cog Pound	#33 Super Heavyweight		2/1 (66.67%)
Junkyard Dog v2.0	Team Dog House	#36 Lightweight	3 steel spikes	3/2 (60.00%)
Junkyard Offspring	Junkyard Dogs	#24 Heavyweight	Punching spikes	3/2 (60.00%)
K.I.S.S.	Team K.I.S.S.	#108 Heavyweight		0/2 (0.00%)
Kegerator	Kegerator	#120 Middleweight	Well-built drive train with a good driver	0/1 (0.00%)
Kegger	Poor College Guys	#77 Middleweight	High-speed bashing arm; stationary impaling spikes	1/1 (50.00%)
Kenny's Revenge	Revenge	#53 Super Heavyweight	Saw	1/1 (50.00%)
Kill-9	Chaos Bots	#127 Middleweight	Plow and I-beam	0/1 (0.00%)
The KillDozer	Team Killdozer	#45 Super Heavyweight	Spiked bulldozer blade	1/2 (33.33%)
Killer Stash	Killer Stash	#189 Lightweight	Wedge; rotating arm; saw; plow	0/1 (0.00%)
KillerB	Strange BRU	#37 Heavyweight	Speed; agility	2/2 (50.00%)
KillerHurtz	KillerHurtz	#4 Heavyweight	Pneumatically powered ax	12/7 (63.16%)
Killer Wasp	Dynamic Duo	#203 Middleweight	Titanium blade, powered by two 6hp gasoline engines	0/1 (00.00%)
Killjoy	BOTWERX	#148 Lightweight	Lifting wedge nose to flip; ramming with rear of bot	0/1 (0.00%)
Kill-O-Amp	Operation Boilermaker	#33 Heavyweight	Kinetic energy	3/6 (33.33%)
Kinetic Kill Vehicle	Hit2Kill	#172 Middleweight	Wedge	0/1 (0.00%)
Knee Breaker	Team Coolrobots	#21 Middleweight		4/4 (50.00%)
Knome	Team Duct Tape	#157 Lightweight	Hammer	0/2 (0.00%)
Knome II	Team Duct Tape	#122 Heavyweight	Blade	0/1 (0.00%)

TT	TM	KOs	KO%	AKOT	KOA	KOA%	AKOAT	TJD	JPA	AA	DA	SA	Points	Builder	Robot Name
3	5	0	0	0:00:00	2	40	0:01:12	1	11	4	3	4	2	Dennis Millard	Instigator
1	1	0	0	0:00:00	1	100	0:01:27	0	0	0	0	0	0	Ken Gentry	Interceptor
1	1	0	0	0:00:00	1	100	0:01:15	0	0	0	0	0	0	Doug Jones	Internal Audit
1	1	0	0	0:00:00	0	0	0:00:00	1	18	7	5	6	0	Ted Shimoda	InTriVerter
1	1	0	0	0:00:00	1	100	0:02:03	0	0	0	0	0	0	Terukazu Nishimura	Iron Eagle
2	3	0	0	0:00:00	0	0	0:00:00	3	18.67	7	6	5.67	2	Jason Demerski	Iron Soldier
1	1	0	0	0:00:00	1	100	0:01:35	0	0	0	0	0	0	Eric Koss	Isosceles 1
1	3	2	66.66	0:01:10	1	33.33	0:01:40	0	0	0	0	0	4	Robert MacDonald	I-Will-Smack-U
1	4	3	75	0:01:18	0	0	0:00:00	1	4	1	1	2	6	Clyde Ward	Jabberwock
1	2	0	0	0:00:00	0	0	0:00:00	1	30	11	11	8	2	Steve Fikar	Jack the Flipper
1	1	0	0	0:00:00	1	100	0:00:45	0	0	0	0	0	0	Anthony King	Jack The Gripper
1	1	0	0	0:00:00	1	100	0:01:25	0	0	0	0	0	0	David Robison	Jack Tripper, The
1	2	0	0	0:00:00	0	0	0:00:00	2	28	9	10	9	2	Allen Pero	Janus
2	4	1	25	0:01:17	1	25	0:00:46	2	22	6.5	6.5	9	4	Curt Meyers	Jaws of Death
1	1	0	0	0:00:00	0	0	0:00:00	1	15	4	5	6	0	Eric Moses	Jersey Devil
2	4	0	0	0:00:00	1	25	0:02:26	3	26.67	9.67	8.33	8.67	6	Norman Muzzy	Jo Mama
1	1	0	0	0:00:00	1	100	0:01:02	0	0	0	0	0	0	John Olson	Jolly Roger
2	6	3	50	0:01:06	2	33.33	0:01:43	1	38	12	14	12	10	Jascha Little	Judge, The
1	2	1	50	0:00:33	0	0	0:00:00	1	17	5	5	7	2	Mike Morrow	JuggerBot
1	4	1	25	0:01:24	1	25	0:00:27	0	0	0	0	0	3	Jascha Little	Juggernaut
1	6	1	16.66	0:03:57	1	16.66	0:02:34	0	0	0	0	0	5	Jim Smentowski	Junior
1	3	1	33.33	0:01:14	0	0	0:00:00	1	8	2	4	2	4	Jerame Powell	Junkyard Cog, The
2	5	2	40	0:02:21	1	20	0:00:59	1	21	5	6	10	6	Kenneth Demshki	Junkyard Dog v2.0
2	5	2	40	0:01:22	1	20	0:01:37	2	15.5	5.5	5	5	6	Anthony Elder	Junkyard Offspring
1	2	0	0	0:00:00	2	100	0:01:24	0	0	0	0	0	0	Steven Nelson	K.I.S.S.
1	1	0	0	0:00:00	0	0	0:00:00	1	18	6	6	6	0		Kegerator
1	2	1	50	0:01:06	1	50	0:02:59	0	0	0	0	0	2	Nathan Roseborrough	Kegger
1	2	0	0	0:00:00	0	0	0:00:00	2	32.5	10	12	10.5	2	Thomas Kohlert	Kenny's Revenge
1	1	0	0	0:00:00	0	0	0:00:00	1	13	4	5	4	0	Bob Proctor	Kill-9
2	3	1	33.33	0:00:30	0	0	0:00:00	2	15	6.5	6	2.5	2	Michael Brouillard	KillDozer, The
1	1	0	0	0:00:00	1	100	0:00:36	0	0	0	0	0	0	Jason Hodge	Killer Stash
2	4	1	25	0:00:52	1	25	0:02:24	1	40	13	13	14	4	Mark Kottlowski	KillerB
6	19	7	36.84	0:01:30	2	10.52	0:00:59	6	15	5.5	4.17	5.33	24	John Reid	KillerHurtz
1	0	0	0	0:00:00	0	0	0:00:00	0	0	0	0	0	0	Doreen Deere	Killer Wasp
1	1	0	0	0:00:00	0	0	0:00:00	1	11	2	5	4	0	Bob Knissel	Killjoy
5	9	3	33.33	0:01:03	4	44.44	0:01:39	2	10	5	2.5	2.5	5	Curt Meyers	Kill-O-Amp
1	1	0	0	0:00:00	1	100	0:01:08	0	0	0	0	0	0	Jason Gates	Kinetic Kill Vehicle
2	8	4	50	0:02:17	1	12.5	0:02:03	0	0	0	0	0	8	Jessica Carlberg	Knee Breaker
1	2	0	0	0:00:00	2	100	0:00:51	0	0	0	0	0	0	Jerome Miles	Knome
1	1	0	0	0:00:00	1	100	0:01:03	0	0	0	0	0	0	Jerome Miles	Knome II

Robot Name	Team	Ranking	Weapon	W/L (W/L%)
Kraken	Team Zeus	#107 Middleweight	Vertical splitting wedge	1/1 (50.00%)
Kritical Mass II	Team Kritical Mass	#45 Heavyweight	Spinning Blade O' Death	2/1 (66.67%)
Lambo	Born2bot	#166 Lightweight	Kinetic energy	0/1 (0.00%)
Land Mine	Team TNT2	#92 Lightweight	Spiked ramming surface; protruding/retracting poker	1/2 (33.33%)
Learning Curve	KE	#102 Lightweight	Pushing; ramming	1/1 (50.00%)
Leper Messiah	Certain Death	#123 Heavyweight	Spinner	0/1 (0.00%)
License to Kill	S.P.E.C.T.R.E. Robotics	#102 Heavyweight	Wedge; mace tail	0/2 (0.00%)
Li'l Chipper	Team Playa	#73 Middleweight	Spinning horizontal blade	1/2 (33.33%)
Little Blue Engine	Team Circuit Breaker	#10 Super Heavyweight		5/2 (71.43%)
Little Drummer Boy	Team Dangerous Drum	#4 Middleweight	25-lb. 4000 RPM spinning steel drum	7/2 (77.78%)
Little Egypt	Sphinxters	#63 Super Heavyweight	4 spinning hammers	1/2 (33.33%)
The Little Engine that Killed	Team Shenanigans	#64 Lightweight	Interchangeable ramming bumper system; hinged wedge with grabbing teeth	2/1 (66.67%)
Little Hater	Killer BotZ	#101 Lightweight	Pneumatic L-shaped lever	1/1 (50.00%)
Little Piece of Hate	Team Violence	#190 Lightweight	Spinning drum	0/1 (0.00%)
Little Pharma	Pharmapac	#36 Middleweight	Wedge	3/1 (75.00%)
Little Punch	Green Iguana	#139 Lightweight	Large steel spike	0/1 (0.00%)
Little Sister	Team Big Brother	#18 Heavyweight	CO_2-powered ram flipper	3/2 (60.00%)
Little Slice	Team Coolrobots	#159 Middleweight		0/2 (0.00%)
Little Wicked	Team Diablo	#207 Lightweight	Large menacing jaws that can wedge under and flip or capture and throw	0/1 (0.00%)
Lobotomizer	G|P|G	#160 Lightweight	Spear	0/2 (0.00%)
Lobotomy	Team GrayMatter	#64 Middleweight	Spinning disk with flails	2/1 (66.67%)
Lock-Nut Monster	Lock-Nut Monster	#115 Lightweight		1/2 (33.33%)
The Locksmith	Five Thieves and a Locksmith	#101 Super Heavyweight	40-lb. steel blades tipped with 5-lb. locks spinning at 600 MPH	0/1 (0.00%)
Locomotion	Silicon Valley Destruction Company	#28 Lightweight	Tenacity; speed	4/2 (66.67%)
Low Blow	Basher of Bots	#30 Lightweight		4/1 (80.00%)
Lug Nuts	Balls of Steel	#140 Heavyweight	Torque	0/1 (0.00%)
M.I.A. (Missing In Action)	Team Black Op Six	#173 Middleweight	Wedge; speed	0/1 (0.00%)
M.O.E. (Marvel of Engineering)	Team Flaming Monkeys	#14 Heavyweight	Spinning ditch cutter	5/3 (62.50%)
Machiavelli	Fear	#174 Middleweight	Dual saw blades	0/1 (0.00%)
Mad Cow	BunsenTech	#191 Lightweight	Spinning dome	0/1 (0.00%)
Maddgoth	Team Crash	#74 Middleweight	Wedge; ram	1/2 (33.33%)
Maggot	Team Maggot	#62 Middleweight	Lifting arm	2/2 (50.00%)
Malice	Team Violence	#84 Lightweight	Wedge; scoop; saw; hooks	1/1 (50.00%)
Malicious Mischief	Team Malicious	#34 Heavyweight	Spring-loaded sledgehammer	3/3 (50.00%)
Malvolio	Raybotics	#32 Middleweight	1000 RPM spinning bar with 1-lb. weights on the ends	3/2 (60.00%)
Maneater	Brooklyn Powers	#116 Middleweight	Wedge with spring-loaded internal hook	0/1 (0.00%)
Manic Aggressive 2.0	Team Final Jeopardy	#61 Lightweight	Pneumatic spike; wedge	2/2 (50.00%)
MantaRay	Team Ray	#73 Super Heavyweight	Dual 10-in. circular carbide blades; 20-in. spring steel blade; air hammer	0/2 (0.00%)

TT	TM	KOs	KO%	AKOT	KOA	KOA%	AKOAT	TJD	JPA	AA	DA	SA	Points	Builder	Robot Name
1	2	0	0	0:00:00	1	50	0:00:43	0	0	0	0	0	2	James Moore	Kraken
1	3	0	0	0:00:00	0	0	0:00:00	3	31.67	10.67	10.67	10.33	4	Jeff Cesnik	Kritical Mass II
1	1	0	0	0:00:00	0	0	0:00:00	1	0	0	0	0	0	Andrew Gardner	Lambo
2	3	0	0	0:00:00	0	0	0:00:00	3	21	7	6.33	7.67	2	Mark Talcott	Land Mine
1	2	0	0	0:00:00	0	0	0:00:00	2	20	6	6.5	7.5	2	Tim Cristy	Learning Curve
1	1	0	0	0:00:00	1	100	0:01:41	0	0	0	0	0	0	Eric DeMoney	Leper Messiah
2	2	0	0	0:00:00	1	50	0:01:45	1	4	1	2	1	0	Valek Sykes	License To Kill
2	3	1	33.33	0:00:34	1	33.33	0:01:14	1	19	6	7	6	2	Scott Martin	Li'l Chipper
2	7	3	50	0:01:20	0	0	0:00:00	3	21.67	7.67	7	7	8	Josh O'Briant	Little Blue Engine
2	9	5	55.55	0:01:40	1	11.11	0:02:54	3	27	8.67	10.67	7.67	15	Steve Buescher	Little Drummer Boy
2	3	0	0	0:00:00	1	33.33	0:01:48	1	6	3	1	2	2	Jim Kennon	Little Egypt
1	3	0	0	0:00:00	0	0	0:00:00	2	21.5	10	8	3.5	4	James Price	Little Engine that Killed, The
1	2	0	0	0:00:00	0	0	0:00:00	2	22.5	7.5	7	8	2	Mark Kottlowski	Little Hater
1	1	0	0	0:00:00	1	100	0:01:51	0	0	0	0	0	0	Lionel Vogt	Little Peice of Hate
1	4	1	25	0:00:56	1	25	0:00:43	2	33	11	9.5	12.5	6	Brian Glasz	Little Pharma
1	1	0	0	0:00:00	0	0	0:00:00	1	15	4	4	7	0	Thomas Beaver	Little Punch
2	5	2	40	0:01:53	2	40	0:01:44	1	43	14	15	14	8	Ian Watts	Little Sister
1	2	0	0	0:00:00	1	50	0:03:57	0	0	0	0	0	0	Christian Carlberg	Little Slice
1	0	0	0	0:00:00	0	0	0:00:00	0	0	0	0	0	0	Zach Bieber	Little Wicked
2	1	0	0	0:00:00	1	100	0:01:37	0	0	0	0	0	0	Michael Glenn	Lobotomizer
1	3	0	0	0:00:00	0	0	0:00:00	3	30.67	9.67	11	10	4	Israel Mathewson	Lobotomy
1	3	1	33.33	0:03:32	0	0	0:00:00	0	0	0	0	0	1	Ben Wright	Lock-Nut Monster
1	1	0	0	0:00:00	1	100	0:00:42	0	0	0	0	0	0	Jason Strohm	Locksmith, The
2	6	1	16.66	0:01:48	0	0	0:00:00	5	20.8	7.4	6.2	7.2	8	David Hall	Locomotion
1	5	1	25	0:04:57	0	0	0:00:00	2	18.5	6	7	5.5	6	Wray Russ	Low Blow
1	0	0	0	0:00:00	0	0	0:00:00	0	0	0	0	0	0	Shawn Ouderkirk	Lug Nuts
1	1	0	0	0:00:00	1	100	0:01:46	0	0	0	0	0	0	Steven West	M.I.A. (Missing In Action)
3	8	2	25	0:01:32	1	12.5	0:02:22	5	30.4	10.2	9.8	10.4	10	Norman Muzzy	M.O.E. (Marvel of Engineering)
1	1	0	0	0:00:00	1	50	0:00:58	0	0	0	0	0	2	Christo Logan	Machiavelli
1	1	0	0	0:00:00	1	100	0:01:44	0	0	0	0	0	0	Justin Morgenthau	Mad Cow
2	3	1	33.33	0:00:44	0	0	0:00:00	1	5	2	1	2	2	William Bottenberg	Maddgoth
2	4	0	0	0:00:00	0	0	0:00:00	4	25	9	7.75	8.25	4	Gary Warren	Maggot
1	2	1	50	0:01:36	1	50	0:01:00	0	0	0	0	0	2	Lionel Vogt	Malice
2	6	1	16.66	0:01:24	1	16.66	0:01:24	1	10	0	5	5	5	Daniel Rupert	Malicious Mischief
2	5	3	60	0:01:30	1	20	0:02:31	1	4	2	1	1	6	Ramon Ebert	Malvolio
1	1	0	0	0:00:00	0	0	0:00:00	1	20	7	6	7	0	Amy Sperber	Maneater
2	4	12.0	25	0:02:00	1	25	0:00:45	2	21	7	7.5	6.5	4	Eric Sporer	Manic Aggressive
2	2	0	0	0:00:00	1	50	0:00:56	1	18	7	6	5	0	Ted Shimoda	MantaRay

Robot Name	Team	Ranking	Weapon	W/L (W/L%)
Mantis	Spike	#55 Lightweight	Linear-actuated crushing jaw	2/1 (66.67%)
Marauder	Team Minus Zero	#32 Super Heavyweight	Bot Stinger	2/2 (50.00%)
Massacre	Team Malicious	#124 Heavyweight		0/1 (0.00%)
The Master	Team Sinister	#44 Middleweight	Gas-powered cutoff saw; actuated lifting arm	3/1 (75.00%)
The Matador	Inertia Labs	#26 Heavyweight	CO_2-powered flipping arm	3/1 (75.00%)
Matts Bammer	Fast Electric Robots	#40 Lightweight	Ramming power; lifting arm	3/2 (60.00%)
Mauler 51-50		#35 Heavyweight	Spinning steel axes; flippers	2/6 (25.00%)
Maximum Overdrive	Team Overdrive	#125 Heavyweight	Wedge with turbo lift	0/1 (0.00%)
Maximum Paralysis	Chaos	#95 Heavyweight	Pneumatically operated composite flippers	0/1 (0.00%)
Maximus	Team Gladiator	#40 Super Heavyweight	Wedge; 14-in. spear; knife blades	2/2 (50.00%)
Mechadon	Team Sinister	#36 Super Heavyweight	Steel claws	2/2 (50.00%)
Mechanical Wildcat	ASME Wildcats	#175 Middleweight	Wedge; 2600 RPM spinning disk	0/1 (0.00%)
MechaVore	Team Shrapnel	#13 Heavyweight	33-lb. flywheel rotating at 2000 RPM; 3-in. stainless steel teeth	5/2 (71.43%)
Meltdown	Bots of Wrath	#128 Middleweight	24-in. steel spinner disk	0/1 (0.00%)
Metalhead	Team Metalhead	#80 Super Heavyweight	6-ft. harpoon	0/1 (0.00%)
Metallic Mayhem	Team Malicious	#110 Middleweight	Wedge	0/2 (0.00%)
MicroVore	Shrapnel	#176 Middleweight	12-lb. 2500 RPM flywheel with two 3-lb. teeth; lifting apparatus	0/1 (0.00%)
The Midnight Project	Team Midnight	#192 Lightweight		0/1 (0.00%)
Minion	Team Coolrobots	#1 Super Heavyweight	Saw blade	11/3 (78.57%)
Minnesota Mangler	Invisible Works	#138 Middleweight	1500 RPM rotational inertia	0/1 (0.00%)
Minotaur	Maurer Power	#111 Heavyweight	Very powerful arm	0/1 (0.00%)
Miriah	Robo Supply	#177 Middleweight	Windmill arm with a chrome ball at each end	0/1 (0.00%)
Misanthrobot X2	Team Misanthrobotics	#155 Lightweight	Spinning saw-like blade	0/2 (0.00%)
The Missing Link	The Infernolab	#49 Lightweight	Gas-powered chainsaw with carbide chain; 2 spear gun projectiles; 3-ft. pointed stick; 2 hardened steel pokers	2/3 (40.00%)
Misty the WonderBot	Robot Action League	#39 Middleweight	22-lb. 4500 RPM horizontal/vertical flywheel	3/1 (75.00%)
Mjollnir	Kampfgruppe5	#55 Heavyweight	Ax	1/4 (20.00%)
Mobster	MAG	#155 Middleweight	Body	0/2 (0.00%)
Moebius	UCF Rc^2	#108 Middleweight	Umbrella spinner	1/1 (50.00%)
Mongus	Team LungFish	#102 Middleweight	Pickax legs; vertical spinning axes	1/1 (50.00%)
Monster	RKDx2	#53 Heavyweight	Spike	1/4 (20.00%)
Mordicus	Team USAFA01	#143 Middleweight	Twin counter-rotating saw blades	0/1 (0.00%)
Mortis	Random Violence Technology	#32 Heavyweight	Sledgehammer-style piercing blade; flipping arm	3/2 (60.00%)
Mouser Classic	Team Catbot	#9 Lightweight		7/6 (53.85%)
Mr. Bonestripper	Team Wetware	#141 Heavyweight	.5 x 4.25 x 48″ titanium blade	0/1 (0.00%)
Mr. Roboto	Kiil Over	#131 Middleweight	Kinetic energy	0/1 (0.00%)
Mr. Snoogglepuss	Scroungers	#134 Lightweight	Spinning rotor with chains and flailing grappling hooks	0/1 (0.00%)
Mr. Sucko	Team Destro	#178 Middleweight	Wedge	0/1 (0.00%)
Musashi	Team Katana	#79 Middleweight	Dual spinning disks in front; wedge in rear	1/1 (50.00%)
My Son	Team Paragon Computers	#52 Heavyweight	Dual reciprocating spike	1/1 (50.00%)

TT	TM	KOs	KO%	AKOT	KOA	KOA%	AKOAT	TJD	JPA	AA	DA	SA	Points	Builder	Robot Name
1	3	1	33.33	0:01:08	1	33.33	0:02:00	1	36	12	13	11	4	Jonathan Bayles	Mantis
2	4	1	25	0:01:11	1	25	0:01:14	2	22.5	7	7	8.5	4	Patrick Campbell	Marauder
1	1	0	0	0:00:00	1	100	0:01:24	0	0	0	0	0	0	Daniel Rupert	Massacre
1	3	0	0	0:00:00	0	0	0:00:00	2	32	11.5	10	10.5	6	Mark Setrakian	Master, The
1	4	2	50	0:01:29	1	25	0:01:42	1	25	7	8	10	6	Alexander Rose	Matador, The
2	5	1	20	0:02:42	2	40	0:01:59	2	37.5	14	13	10.5	6	Jeff Vasquez	Matts Bammer
5	8	2	25	0:00:47	3	37.5	0:01:38	1	20	6	7	7	4	Charles Tilford	Mauler 51-50
1	1	0	0	0:00:00	1	100	0:01:15	0	0	0	0	0	0	Jeff Mills	Maximum Overdrive
1	1	0	0	0:00:00	0	0	0:00:00	1	13	6	7	0	0	Charlie King	Maximum Paralysis
2	4	0	0	0:00:00	0	0	0:00:00	4	19.25	7	6.5	5.75	4	Patrick Lindstrom	Maximus
2	3	1	33.33	0:01:55	0	0	0:00:00	1	5	0	0	5	4	Mark Setrakian	Mechadon
1	1	0	0	0:00:00	1	100	0:01:43	0	0	0	0	0	0	Andrew McCutchen	Mechanical Wildcat
2	7	2	28.57	0:01:24	1	14.28	0:02:37	3	21	7.33	7	6.67	10	Robert L. Marzinske	MechaVore
1	1	0	0	0:00:00	0	0	0:00:00	1	13	5	3	5	0	Tom Wickey	Meltdown
1	1	0	0	0:00:00	0	0	0:00:00	1	13	2	7	4	0	John Bamberg	Metalhead
2	2	0	0	0:00:00	0	0	0:00:00	2	14	7	4.5	2.5	0	Daniel Rupert	Metallic Mayhem
1	1	0	0	0:00:00	1	100	0:00:51	0	0	0	0	0	0		MicroVore
1	1	0	0	0:00:00	1	100	0:01:20	0	0	0	0	0	0	Ben Wright	Midnight Project, The
5	14	3	21.42	0:01:05	2	14.28	0:01:15	9	30.33	11.56	9.22	9.56	29	Christian Carlberg	Minion
1	1	0	0	0:00:00	0	0	0:00:00	1	9	1	1	7	0	Charles Habermann	Minnesota Mangler
1	1	0	0	0:00:00	0	0	0:00:00	1	0	0	0	0	0	David Maurer	Minotaur
1	1	0	0	0:00:00	1	100	0:00:41	0	0	0	0	0	0	Jack Harwick	Miriah
2	2	0	0	0:00:00	2	100	0:01:29	0	0	0	0	0	0	Erik Reikes	Misanthrobot X2
2	5	2	40	0:01:29	3	60	0:02:28	0	0	0	0	0	4	Jason Bardis	Missing Link, The
1	4	1	25	0:01:20	0	0	0:00:00	3	23	7	6.67	9.33	6	Will Wright	Misty the WonderBot
4	4	1	25	0:00:52	1	25	0:02:47	2	5.5	3	0.5	2	2	Mike Flanagan	Mjollnir
2	2	0	0	0:00:00	2	100	0:01:17	0	0	0	0	0	0	Larry Magnoli	Mobster
1	2	0	0	0:00:00	1	50	0:00:53	0	0	0	0	0	2	Brett Dawson	Moebius
1	2	0	0	0:00:00	0	0	0:00:00	1	16	3	6	7	2	Kevin Lung	Mongus
3	5	1	20	0:00:33	2	40	0:03:02	1	15	5	10	0	2	Kevin Knoedler	Monster
1	1	0	0	0:00:00	0	0	0:00:00	1	7	1	3	3	0	Barry Mullins	Mordicus
2	5	0	0	0:00:00	1	20	0:02:10	3	25	10	8.33	6.67	6	Rob Knight	Mortis
5	13	3	23.07	0:02:51	4	30.76	0:02:49	5	25.4	5.6	10.2	9.6	14	Fon Davis	Mouser Classic
1	0	0	0	0:00:00	0	0	0:00:00	0	0	0	0	0	0	Ray Scully	Mr. Bonestripper
1	1	0	0	0:00:00	0	0	0:00:00	1	12	4	3	5	0	Kyle Hickerson	Mr. Roboto
1	1	0	0	0:00:00	0	0	0:00:00	1	17	7	6	4	0	Jonathan Stoffer	Mr. Snoogglepuss
1	1	0	0	0:00:00	1	100	0:01:23	0	0	0	0	0	0	Doug Welch	Mr. Sucko
1	2	1	50	0:01:40	0	0	0:00:00	1	11	4	3	4	2	Evan White	Musashi
1	2	1	50	0:00:31	1	50	0:01:11	0	0	0	0	0	2	Keith Gresham	My Son

Robot Name	Team	Ranking	Weapon	W/L (W/L%)
Namreko 3000	Gammatronic Robot Brigade	#69 Heavyweight	14-in. rescue saw blades	1/3 (25.00%)
Narwhal	The Wrecking Crew	#133 Middleweight	Flipper/thrower	0/1 (0.00%)
Nasty Overbite	Team Jawbreaker	#62 Heavyweight	Large clamping jaw	1/1 (50.00%)
Neb	Nebecannezer	#179 Middleweight	Rotating saws on a wedge	0/1 (0.00%)
Necronomibot	Team Misanthrobotics	#180 Middleweight	Flywheel actuated stabbers	0/1 (0.00%)
Negative Reinforcement	Hunter Robotics	#96 Heavyweight	Pneumatic claw	0/1 (0.00%)
Nemisis	Silicon Valley Destruction Company	#109 Heavyweight	Speed	0/2 (0.00%)
Neptune	Team Stingray	#102 Super Heavyweight	Array of 6 pneumatic cannons	0/1 (0.00%)
Nerd Killer	Team SMC	#83 Heavyweight	Lifting arm	0/1 (0.00%)
New Cruelty	Killerbotics	#6 Super Heavyweight	22-lb. I-beam ram	6/1 (85.71%)
Nibbler	Team S.I.U. Robots	#204 Middleweight	Pincers	0/1 (0.00%)
Nicebot	Team Nice Guys	#193 Lightweight	3-ft. spinning drill rod with 2 sledgehammer heads	0/1 (0.00%)
Nightmare	Team Nightmare	#9 Heavyweight	Spinning disk	8/7 (53.33%)
No Apologies	Tork Team	#27 Super Heavyweight	Spike	2/2 (50.00%)
No Tolerance II	No Tolerance Combat Robotics	#89 Middleweight	ClampBot	1/1 (50.00%)
No Tolerance III	No Tolerance Combat Robotics	#31 Lightweight	Robo-lift	3/3 (50.00%)
None Shall Pass	Webtrends	#144 Lightweight	0.5-in. spinning titanium bar	0/1 (0.00%)
Nsyncerator	Team Tatar	#161 Lightweight	20-lb. 3360 RPM spinning blade	0/2 (0.00%)
Odin II	Team Odin	#20 Super Heavyweight	75 lbs. of spinning steel	3/4 (42.86%)
Ogre	Team Ogre	#18 Super Heavyweight	7 spikes	4/2 (66.67%)
Omega-13	Regan Designs	#31 Heavyweight	Inertia	3/2 (60.00%)
One Tin Soldier	One Tin Soldier	#90 Super Heavyweight	Carbide noogie prongs	0/1 (0.00%)
Orion	Pathfinder	#126 Heavyweight	Dual spring-loaded 8-in. pickaxes on multi-positional side turrets	0/1 (0.00%)
The Orlando Commando	Team O-Town	#46 Middleweight		2/1 (66.67%)
Outlaw	Designated Destruction	#85 Middleweight	Ramming	1/2 (33.33%)
Overbearing	Mongoose	#64 Heavyweight	60-lb. spinning mass with a modular blade system	1/1 (50.00%)
OverKill	Team Coolrobots	#5 Heavyweight	Super-big Rambo blade	10/4 (71.43%)
Overlord	Pierce	#81 Super Heavyweight	Power and (maybe) a lifting arm	0/1 (0.00%)
Pack Raptors	Team Raptor	#31 Middleweight	Lifter arms	4/2 (66.67%)
Paladin	Team Batteraizees	#135 Lightweight	Hammer	0/1 (0.00%)
Palindrome 2.0	Team Radicus	#156 Middleweight	Dual metal circular saws	0/2 (0.00%)
Panic Attack 3 (PA3)	Team Panic Attack	#59 Heavyweight	Electric-powered lifting spikes	1/1 (50.00%)
Parallax	Team Attack	#140 Lightweight	Spinning mace balls	0/1 (0.00%)
Paranoid Android	Paranoid Android	#90 Lightweight		1/2 (33.33%)
Parasitic	Team E C	#76 Heavyweight	Front lifting arm with interchangeable weapons	1/1 (50.00%)
Patriot	Team Bomb	#87 Lightweight	Wedge; flipper	1/1 (50.00%)
Patton	Team Patton	#79 Super Heavyweight	Pneumatic ram	0/1 (0.00%)
Pegleg	Team Deranged	#181 Middleweight	Luck	0/1 (0.00%)
Pestilence	Team Revelation	#89 Lightweight	Pneumatic lifter	1/2 (33.33%)
Pharmapac	Pharmapac	#38 Super Heavyweight	Pneumatic flipping wedge	2/1 (66.67%)

TT	TM	KOs	KO%	AKOT	KOA	KOA%	AKOAT	TJD	JPA	AA	DA	SA	Points	Builder	Robot Name
2	4	0	0	0:00:00	2	50	0:01:32	1	45	15	15	15	2	Mike Okerman	Namreko 3000
1	1	0	0	0:00:00	0	0	0:00:00	1	10	3	4	3	0	Scott Kilbourne	Narwhal
1	2	1	50	0:01:21	0	0	0:00:00	1	21	6	6	9	2	Jeff Harth	Nasty Overbite
1	1	0	0	0:00:00	1	100	0:01:16	0	0	0	0	0	0	Karen Sharp	Neb
1	1	0	0	0:00:00	1	100	0:01:00	0	0	0	0	0	0	Erik Reikes	Necronomibot
1	1	0	0	0:00:00	0	0	0:00:00	1	12	4	4	4	0	Glen Sears	Negative Reinforcement
2	1	0	0	0:00:00	1	100	0:00:45	0	0	0	0	0	0	David Hall	Nemisis
1	1	0	0	0:00:00	1	100	0:00:33	0	0	0	0	0	0	Simon Arthur	Neptune
1	1	0	0	0:00:00	0	0	0:00:00	1	22	6	7	9	0	Casey Bennett	Nerd Killer
1	7	2	28.57	0:01:07	0	0	0:00:00	5	24.2	8.2	7.4	8.6	13	Richard Stuplich	New Cruelty
1	0	0	0	0:00:00	0	0	0:00:00	0	0	0	0	0	0	AM SE	Nibbler
1	1	0	0	0:00:00	1	100	0:01:47	0	0	0	0	0	0	Jason Crum	Nicebot
6	15	4	26.66	0:01:42	3	20	0:01:03	5	21	5.8	8	7.2	15	Jim Smentowski	Nightmare
2	4	2	50	0:01:29	1	25	0:01:54	1	22	7	8	7	4	John E. Torkelson	No Apologies
1	2	0	0	0:00:00	0	0	0:00:00	2	26	8.5	7.5	10	2	Stefan Nock	No Tolerance II
3	6	3	50	0:02:11	0	0	0:00:00	3	7.33	2.67	2	2.67	6	Stefan Nock	No Tolerance III
1	1	0	0	0:00:00	0	0	0:00:00	1	12	3	4	5	0	James Bean	None Shall Pass
2	1	0	0	0:00:00	1	100	0:01:50	0	0	0	0	0	0	Eric Stoliker	Nsyncerator
4	7	1	14.28	0:01:03	0	0	0:00:00	5	17.8	6	6.6	5.2	7	John McKenzie	Odin II
2	6	0	0	0:00:00	0	0	0:00:00	6	26.5	9	9	8.5	8	Tom Griffin	Ogre
2	5	0	0	0:00:00	0	0	0:00:00	5	24.6	8.6	8.8	7.2	6	Brent Regan	Omega-13
1	1	0	0	0:00:00	0	0	0:00:00	1	0	0	0	0	0	Dane Scarborough	One Tin Soldier
1	1	0	0	0:00:00	1	100	0:01:43	0	0	0	0	0	0	Christopher Hedberg	Orion
1	3	2	66.66	0:01:03	0	0	0:00:00	1	11	3	3	5	4	Robert Wilburn	Orlando Commando, The
2	3	0	0	0:00:00	0	0	0:00:00	3	25.33	7.67	8.67	9	2	Mike Rede	Outlaw
1	1	1	100	0:01:27	0	0	0:00:00	0	0	0	0	0	2	Lucas Gies	Overbearing
4	14	4	28.57	0:01:35	0	0	0:00:00	10	22.2	7.7	7.4	7.1	21	Christian Carlberg	OverKill
1	1	0	0	0:00:00	0	0	0:00:00	1	13	5	4	4	0	Ronnie Pierce	Overlord
2	6	0	0	0:00:00	1	16.66	0:01:10	5	27.6	9.6	8.2	9.8	8	Chuck Pitzer	Pack Raptors
1	1	0	0	0:00:00	0	0	0:00:00	1	17	11	5	1	0	Justin Chou	Paladin
2	2	0	0	0:00:00	2	100	0:01:08	0	0	0	0	0	0	Anthony Hall	Palindrome 2.0
1	2	1	50	0:01:10	1	50	0:02:07	0	0	0	0	0	2	Kim Davies	Panic Attack 3 (PA3)
1	1	0	0	0:00:00	0	0	0:00:00	1	15	5	5	5	0		Parallax
1	3	1	33.33	0:03:09	2	66.66	0:03:43	0	0	0	0	0	2	Dan Haeg	Paranoid Android
1	2	0	0	0:00:00	1	50	0:02:20	1	27	10	8	9	2	Jon Orrell	Parasitic
1	2	1	50	0:01:48	0	0	0:00:00	1	19	7	5	7	2	George Roach	Patriot
1	1	0	0	0:00:00	0	0	0:00:00	1	15	6	7	2	0	Lester Lilley	Patton
1	1	0	0	0:00:00	1	100	0:00:31	0	0	0	0	0	0	Grayson DuRaine	Pegleg
2	2	1	50	0:02:35	1	50	0:02:40	0	0	0	0	0	2	Todd Mendenhall	Pestilence
1	3	0	0	0:00:00	0	0	0:00:00	3	29.67	9.67	10.67	9.33	4	Brian Glasz	Pharmapac

Robot Name	Team	Ranking	Weapon	W/L (W/L%)
Phere	Team Xtremebots	#17 Super Heavyweight	Spinning dome	4/2 (66.67%)
Phishfeuxd	Team Osoka	#182 Middleweight	Spinning hooks; titanium wedge	0/1 (0.00%)
Phoenix	Team Flaming Monkeys	#115 Middleweight	Pickax; rapid fire	0/2 (0.00%)
Phrizbee	Team LOGICOM	#46 Heavyweight	80-lb. 1000 RPM spinning shell with steel blades	2/2 (50.00%)
The Piecemaker	No Mercy	#127 Heavyweight		0/1 (0.00%)
Pink Slip	Team Severance	#146 Middleweight	4-in. 10,000 RPM vertical cutting disk	0/1 (0.00%)
Playmate	MITGIT	#205 Middleweight		0/1 (0.00%)
Plowbot	C-M Engineering	#53 Middleweight	Lifting plow blade	2/1 (66.67%)
Pokey	Team Sallad	#57 Middleweight	Lifter arm	2/2 (50.00%)
Portable Killsaw	Team BattleBobs	#183 Middleweight	Spinning steel ring with teeth; small hammers	0/1 (0.00%)
Potter's Wheel	GrinElf	#184 Middleweight	Low flywheel	0/1 (0.00%)
Pressure Drop	Automatum	#47 Middleweight		2/1 (66.67%)
Pro-AM	Robotdojo	#69 Super Heavyweight	Lifting/clamping arm; 30-in. hammer	0/1 (0.00%)
The Probe	The Penetrators	#22 Super Heavyweight		3/2 (60.00%)
Problem Child	Team Anthrax	#68 Middleweight	25-lb. 1300 RPM flywheel	2/2 (50.00%)
Project: Y.A.W.N	Team JaGGeD	#58 Lightweight	300 MPH spinner; wedge	2/1 (66.67%)
Prompt Critical	Prompt Critical	#103 Super Heavyweight	Hammer of Fury	0/1 (0.00%)
Proto-Type 4	Black Ops	#91 Super Heavyweight	Cleated horizontal spinning drum	0/1 (0.00%)
Psycho Splatter	Flying Parts	#128 Heavyweight	Twin 32-in. horizontal saw blades	0/1 (0.00%)
Psychotron	Team Rockitz	#30 Middleweight	Spinning wedge design with chrome spikes	4/2 (66.67%)
Punjar	Team Punjar	#8 Heavyweight	Wedge	10/6 (62.50%)
Purple Haze	Alfred University	#194 Lightweight	Kinetic energy; lifting arm	0/1 (0.00%)
Pursuer	Pursuer	#21 Lightweight		5/2 (71.43%)
PyRAMidroid	PyRAMidroid	#50 Heavyweight	Skirt wedge; opposing 12-in. pneumatic spikes	2/2 (50.00%)
RA	Team RA	#185 Middleweight	Kill-saw tail; 2 front kill-saw claws	0/1 (0.00%)
RACC	Robot Action League	#156 Lightweight		0/2 (0.00%)
Rama Lama Ding Bot	Kirwan	#123 Lightweight	Primary weapon is speed	0/1 (0.00%)
Rambite	Robotic Death Company	#25 Lightweight	1250 RPM spinning blades	4/1 (80.00%)
Ramming Speed	Team Clobber	#91 Lightweight	Pusher/rammer	1/2 (33.33%)
Rammstein	Loki Robotics	#7 Super Heavyweight	Pneumatic spike	6/4 (60.00%)
Rampage	Team Rampage	#76 Lightweight	5-in. vertical blade	1/1 (50.00%)
RASTA	Robot Action League	#158 Lightweight		0/2 (0.00%)
Ravager	Ravager	#49 Middleweight	Pneumatic spike hammer	2/2 (50.00%)
Razer	Razer	#21 Heavyweight	Hydraulic squeeze/piercer	3/3 (50.00%)
Reactore	Pilgrim Nuclear Technology Alliance	#129 Heavyweight	Archimedes screw and pusher	0/1 (0.00%)
Reaper	Team Caffeine	#51 Lightweight	Large ax	2/1 (66.67%)
Rebob	New Tech Robotics	#157 Middleweight	2 horizontally swinging titanium pneumatic rock picks	0/2 (0.00%)
Recycler	Team Recycle	#67 Lightweight	Wedge; lifting arm	2/1 (66.67%)
Red Scorpion	Team Scorpion	#148 Middleweight	Stinger	0/3 (0.00%)
Revision Z	Team Malicious	#15 Super Heavyweight	Spiked tail; steel bumper	4/4 (50.00%)

TT	TM	KOs	KO%	AKOT	KOA	KOA%	AKOAT	TJD	JPA	AA	DA	SA	Points	Builder	Robot Name
2	6	1	16.66	0:00:56	2	33.33	0:00:47	3	25	8.33	8.67	8	8	Gaylan Douglas	Phere
1	1	0	0	0:00:00	1	100	0:00:46	0	0	0	0	0	0	Michael Macht	Phishfeuxd
2	2	0	0	0:00:00	1	50	0:01:24	1	21	6	7	8	0	Norman Muzzy	Phoenix
2	4	0	0	0:00:00	1	25	0:02:07	3	25.67	7.67	8.33	9.67	4	Brian Nave	Phrizbee
1	1	0	0	0:00:00	1	100	0:00:31	0	0	0	0	0	0	Mark Nenadic	Piecemaker, The
1	1	0	0	0:00:00	0	0	0:00:00	1	5	2	1	2	0		Pink Slip
1	0	0	0	0:00:00	0	0	0:00:00	0	0	0	0	0	0	William Garcia	Playmate
1	3	1	33.33	0:01:20	0	0	0:00:00	2	20	6	8	6	4	Bill Olson	Plowbot
2	4	1	25	0:01:43	0	0	0:00:00	3	20	7.67	6	6.33	4	Dallas Goecker	Pokey
1	1	0	0	0:00:00	1	100	0:01:45	0	0	0	0	0	0	Robert Hamlet	Portable Killsaw
1	1	0	0	0:00:00	1	100	0:01:44	0	0	0	0	0	0	Norman Elfer	Potter's Wheel
1	3	2	66.66	0:01:50	0	0	0:00:00	1	0	0	0	0	4	Derek Young	Pressure Drop
1	1	0	0	0:00:00	0	0	0:00:00	1	22	12	8	2	0	Mike Konshak	Pro-AM
2	5	1	20	0:01:22	0	0	0:00:00	4	24	7.25	8.5	8.25	6	Jeff Banks	Probe, The
2	4	0	0	0:00:00	1	25	0:00:49	2	28.5	7.5	11.5	9.5	4	Mark Thompson	Problem Child
1	3	1	33.33	0:01:23	1	33.33	0:01:00	0	0	0	0	0	4	Joshua Howard	Project: Y.A.W.N.
1	1	0	0	0:00:00	1	100	0:01:55	0	0	0	0	0	0	Ken Russell	Prompt Critical
1	1	0	0	0:00:00	0	0	0:00:00	0	0	0	0	0	0	Rob Meyer	Proto-Type 4
1	1	0	0	0:00:00	1	100	0:01:03	0	0	0	0	0	0	Joel Remer	Psycho Splatter
2	6	0	0	0:00:00	0	0	0:00:00	6	23.33	8	7.5	7.83	8	Kelly Smith	Psychotron
5	16	3	18.75	0:02:28	1	6.25	0:01:08	6	20.83	7.5	5.83	7.5	16	Ramiro Mallari	Punjar
1	1	0	0	0:00:00	1	100	0:01:03	0	0	0	0	0	0	Nick Belton	Purple Haze
2	7	0	0	0:00:00	0	0	0:00:00	5	25.6	9	8.2	8.4	10	Darren Mackenzie	Pursuer
2	4	0	0	0:00:00	1	25	0:01:23	2	20	6.5	6	7.5	4	Mike Konshak	PyRAMidroid
1	1	0	0	0:00:00	1	100	0:02:59	0	0	0	0	0	0	Solomon Spector	RA
1	2	0	0	0:00:00	0	0	0:00:00	0	0	0	0	0	0	Will Wright	RACC
1	1	0	0	0:00:00	0	0	0:00:00	1	22	8	9	5	0	John Hoffman	Rama Lama Ding Bot
1	5	2	40	0:01:27	0	0	0:00:00	3	26.33	6.67	8.67	11	8	John Mladenik	Rambite
2	3	0	0	0:00:00	0	0	0:00:00	3	23.67	8.67	8.33	6.67	2	Joseph Fieber	Ramming Speed
4	10	3	30	0:02:07	2	20	0:01:56	5	21.6	7.6	7	7	12	Nola Garcia	Rammstein
1	2	1	50	0:01:09	1	50	0:00:55	0	0	0	0	0	2	Troy Tracey	Rampage
1	2	0	0	0:00:00	2	100	0:02:38	0	0	0	0	0	0	Jim Sellers	RASTA
2	4	1	25	0:00:46	2	50	0:01:18	1	25	9	10	6	4	Butch Logston	Ravager
2	6	2	33.33	0:03:20	2	33.33	0:02:45	1	10	0	0	10	7	Ian Lewis	Razer
1	1	0	0	0:00:00	1	100	0:01:47	0	0	0	0	0	0	Michael Bastoni	Reactore
1	3	1	33.33	0:02:24	0	0	0:00:00	2	28	8	11	9	4	Robert Wooden	Reaper
2	2	0	0	0:00:00	2	100	0:00:59	0	0	0	0	0	0	Jason Woods	Rebob
1	3	0	0	0:00:00	1	33.33	0:02:42	1	34	12	10	12	4	Don Jennings	Recycler
2	3	0	0	0:00:00	1	33.33	0:00:17	1	3	1	0	2	0	Shane Washburn	Red Scorpion
4	8	2	25	0:01:43	0	0	0:00:00	6	22.5	7	9	6.5	8	Daniel Rupert	Revision Z

Robot Name	Team	Ranking	Weapon	W/L (W/L%)
Rhino	Inertia Labs	#11 Heavyweight	Pneumatic ram	9/3 (75.00%)
Rhino	Inertia Labs	#71 Super Heavyweight	Pneumatic ram	0/1 (0.00%)
Rhode Hog	Team Clam Bot	#63 Lightweight	Wedge	2/2 (50.00%)
Rim Tin Tin	Team Solo	#128 Lightweight	Plow; 8-in. spike	0/3 (0.00%)
Ripoff	Team Litewav	#92 Super Heavyweight	Dual vertically spinning 28-in. saw blades	0/1 (0.00%)
The Ripper	Team Servo	#85 Heavyweight		0/1 (0.00%)
RipTide	Wetware	#134 Middleweight	Twin 15-in. motorcycle rims; twin drum spinner	0/1 (0.00%)
Road Dot	Team Peas	#208 Lightweight	Reckless abandon	0/1 (0.00%)
Road Rage	Killer BotZ	#44 Super Heavyweight	Turret-mounted hammer	2/1 (66.67%)
Robo Master	Robo Master	#99 Lightweight	Lifting wedge	1/2 (33.33%)
RoboMisDirection	RoboSaga	#167 Lightweight		0/1 (0.00%)
RoboSapien	Team BattleBobs	#38 Heavyweight	Spinning steel ring with wedge flippers and claws	2/1 (66.67%)
Robot X	Team Blaze	#121 Lightweight	The Death Roll of Doom Sai—a Steel Sai about 14" long that spins on the (X) axis projecting out in front of the bot	0/2 (0.00%)
RoboVore	Mackenzie Death Machines, LLC	#49 Heavyweight	Pneumatic spear	2/1 (66.67%)
Rock-Hard	G.B. Engineering	#186 Middleweight	Air-powered metal arm with metal spike; basic wedge	0/1 (0.00%)
Rocky Balbota		#104 Super Heavyweight	Speed and power; ramming plate; detachable scooping front wedge	0/1 (0.00%)
Rolling Blackout	Team Overload	#54 Lightweight	Electric lifting arm	2/2 (50.00%)
Ronin	Team Sinister	#44 Heavyweight	5hp gas engine driving a 20" negative racked, carbide tipped saw	2/3 (40.00%)
Ronin	Team Sinister	#19 Super Heavyweight	20-in. negative racked, carbide-tipped saw	4/4 (50.00%)
Root Canal	Mutant Robots	#118 Middleweight		0/1 (0.00%)
Rott-bot 2000	Team Second Breakdown	#153 Lightweight		0/3 (0.00%)
S.L.A.M.	Team S.L.A.M.	#80 Heavyweight	Rotating drum with teeth	1/2 (33.33%)
S.L.A.M.	Team S.L.A.M.	#87 Super Heavyweight	Rotating drum with teeth	0/3 (0.00%)
S.O.B	SOE	#112 Middleweight	Winch-wound torsion spring with combination bashing/flipping arm	0/1 (0.00%)
SABotage	Team Sabotage	#18 Middleweight	Lifting clamping arm	5/3 (62.50%)
Sallad	Team Sallad	#4 Lightweight	Versatile arm	10/6 (62.50%)
Sawzz-All	Team M-Tech	#187 Middleweight	Horizontal saw and rear wedge	0/1 (0.00%)
Scallywag	Welsh Wizards	#84 Heavyweight	Flipper; lifter; powerful push	0/1 (0.00%)
Scheduled for Destruction	Team Kirwan	#81 Middleweight	Pneumatic puncher	1/1 (50.00%)
Scrap Daddy's Flipskanker	Scrap Daddy	#86 Middleweight	Blade	1/3 (25.00%)
Scrap Daddy's Persistor	Scrap Daddy	#82 Heavyweight	Blades; wedges	0/3 (0.00%)
Scrap Daddy's Surplus	Scrap Daddy	#27 Lightweight	Spinner; wedge	4/3 (57.14%)
Scrap Daddy's Turbulence	Scrap Daddy	#88 Super Heavyweight	Rams; flippers; blades	0/2 (0.00%)
Scrap Metal	Team U-FO	#79 Lightweight	Spinner; kinetic ram	1/1 (50.00%)
Scrap Metal	Team U-FO	#188 Middleweight	Spinner; kinetic ram	0/1 (0.00%)
Scrub	Team Vicious	#98 Lightweight	Flailing spiked tail	1/1 (50.00%)
Serial Box Killer	Minimal Control	#20 Lightweight	Ramming; spike	5/4 (55.56%)
Sgt. Spanky II	Overlord	#106 Heavyweight	2, 16-tooth rings	0/2 (0.00%)
Shaft	Robot Action League	#88 Lightweight	The Shaft: 20 lbs. of hardened steel	1/2 (33.33%)
Shaka	DreamDroid	#94 Heavyweight	Dual electric lifting actuators	0/1 (0.00%)

TT	TM	KOs	KO%	AKOT	KOA	KOA%	AKOAT	TJD	JPA	AA	DA	SA	Points	Builder	Robot Name
2	12	6	50	0:02:48	1	8.33	0:01:32	3	25	10	6.67	8.33	12	Alexander Rose	Rhino
1	1	0	0	0:00:00	0	0	0:00:00	1	20	0	5	15	0	Alexander Rose	Rhino
2	4	0	0	0:00:00	1	25	0:02:42	2	24	8.5	8	7.5	4	Michael Thombs	Rhode Hog
3	3	0	0	0:00:00	2	66.66	0:01:50	1	19	8	6	5	0	Dan Watts	Rim Tin Tin
1	1	0	0	0:00:00	0	0	0:00:00	1	0	0	0	0	0	Christopher Hannold	Ripoff
1	1	0	0	0:00:00	0	0	0:00:00	1	21	6	8	7	0	Brian Ondov	Ripper, The
1	1	0	0	0:00:00	0	0	0:00:00	1	10	3	6	1	0	Ray Scully	RipTide
1	0	0	0	0:00:00	0	0	0:00:00	0	0	0	0	0	0	Jan Peasley	Road Dot
1	3	0	0	0:00:00	1	33.33	0:02:14	0	0	0	0	0	4	William Hubbard	Road Rage
2	3	0	0	0:00:00	0	0	0:00:00	3	15.33	5	5.67	4.67	2	Jim Yeh	Robo Master
1	1	0	0	0:00:00	1	100	0:01:41	0	0	0	0	0	0	Suni Murata	RoboMisDirection
1	3	1	33.33	0:01:02	1	33.33	0:01:36	1	42	14	15	13	4	Robert Hamlet	RoboSapien
2	2	0	0	0:00:00	0	0	0:00:00	2	11	3.5	3	4.5	0	Edward Robinson	Robot X
1	3	0	0	0:00:00	0	0	0:00:00	2	24	9.5	8	6.5	4	Richard Mackenzie	RoboVore
1	1	0	0	0:00:00	1	100	0:01:54	0	0	0	0	0	0	Ellery Penas	Rock-Hard
1	1	0	0	0:00:00	1	100	0:01:17	0	0	0	0	0	0	John Huemme	Rocky Balbota
2	3	1	33.33	0:01:00	0	0	0:00:00	1	27	8	10	9	4	Mike May	Rolling Blackout
2	5	1	20	0:02:12	0	0	0:00:00	1	5	0	5	0	4	Peter Abrahamson	Ronin
4	8	0	0	0:00:00	2	25	0:01:13	5	20.2	4.2	8.6	7.4	8	Peter Abrahamson	Ronin
1	1	0	0	0:00:00	0	0	0:00:00	1	19	8	8	3	0	Donald Hutson	Root Canal
2	3	0	0	0:00:00	1	33.33	0:01:02	0	0	0	0	0	0	Brian Scearce	Rott-bot 2000
1	3	0	0	0:00:00	1	33.33	0:00:55	0	0	0	0	0	2	Lowell Nelson	S.L.A.M.
3	2	0	0	0:00:00	2	100	0:01:52	0	0	0	0	0	0	Lowell Nelson	S.L.A.M.
1	1	0	0	0:00:00	0	0	0:00:00	1	22	8	6	8	0	Sony Online Entertainment	S.O.B.
3	8	1	12.5	0:01:43	2	25	0:01:25	5	29	9	10.8	9.2	10	Matt Sabatino	SABotage
5	16	6	37.5	0:02:29	2	12.5	0:03:12	5	24.2	7.4	9.6	7.2	19	Dallas Goecker	Sallad
1	1	0	0	0:00:00	1	100	0:01:47	0	0	0	0	0	0	Rudolfo "Sonny" Saucedo	Sawzz-All
1	1	0	0	0:00:00	0	0	0:00:00	1	22	8	7	7	0	Kim Davies	Scallywag
1	2	1	50	0:01:41	0	0	0:00:00	1	21	7	7	7	2	Richard Wood	Scheduled for Destruction
3	4	0	0	0:00:00	1	25	0:02:43	3	20.67	6.33	7	7.33	2	Mark Bradford	Scrap Daddy's Flipskanker
3	3	0	0	0:00:00	1	33.33	0:01:18	2	16	6	4.5	5.5	0	Mark Bradford	Scrap Daddy's Persistor
3	7	1	14.28	0:01:41	1	14.28	0:01:13	5	22.4	7.6	6.8	8	8	Mark Bradford	Scrap Daddy's Surplus
2	2	0	0	0:00:00	2	100	0:01:25	0	0	0	0	0	0	Mark Bradford	Scrap Daddy's Turbulence
1	2	1	50	0:01:16	1	50	0:02:27	0	0	0	0	0	2	Robert Masek	Scrap Metal
1	1	0	0	0:00:00	1	100	0:00:37	0	0	0	0	0	0	Robert Masek	Scrap Metal
1	2	0	0	0:00:00	0	0	0:00:00	2	23.5	8	7.5	8	2	Mike Regan	Scrub
4	9	0	0	0:00:00	4	44.44	0:01:46	5	27.6	9.4	9.6	8.6	10	Matthew Garten	Serial Box Killer
2	2	0	0	0:00:00	2	100	0:01:10	0	0	0	0	0	0	Patrick Urschel	Sgt. Spanky II
2	3	1	33.33	0:02:20	0	0	0:00:00	2	17	10.5	3	3.5	2	Rik Winter	Shaft
1	1	0	0	0:00:00	0	0	0:00:00	1	16	10	1	5	0	Walter Martinez	Shaka

Robot Name	Team	Ranking	Weapon	W/L (W/L%)
Shark Byte	Team Shark	#19 Heavyweight	Turret spinner	4/2 (66.67%)
Sharp Tooth	With an I	#84 Super Heavyweight	2, 6-in. hole saws; tethered bowling ball	0/1 (0.00%)
She Fights Like a Girl	Team Sunshine Factory	#111 Lightweight	Drum spinner	1/1 (50.00%)
Shish-ka-bot 1.1	Team Kirwan	#104 Lightweight	Air-powered spike	1/2 (33.33%)
Shockwave	Team SMC	#118 Lightweight	Wedge	0/2 (0.00%)
Short Circuit	Team Short Circuit	#147 Middleweight	Pneumatic lifter	0/1 (0.00%)
Short Order Chef	Team Bus Boys	#37 Middleweight	Flipper arm	3/2 (60.00%)
Shovit	Mad Max	#101 Middleweight	Shovel	1/1 (50.00%)
Shredbot	Dreadbot	#189 Middleweight	Spinning bar	0/1 (0.00%)
Shrike	Team Masada	#23 Lightweight	Pneumatic pickax	4/4 (50.00%)
Silver Bullet	Kiss This	#77 Lightweight	Battering wedge	1/1 (50.00%)
Silverback	Team Death by Monkeys	#43 Heavyweight	Lifting arm	2/1 (66.67%)
Sisyphus	Team Bust-a-Bot	#38 Lightweight	Wedge; plow	3/1 (75.00%)
SkidMark	Team Half-Life Inc.	#34 Lightweight	Spinning tri-foils; air-powered flipper	3/2 (60.00%)
Slam Job	BlackRoot	#15 Heavyweight	Wedge with spike slam hammer	5/2 (71.43%)
Slap 'Em Silly	Fatcats	#19 Lightweight	Front spike; rear fixed wedge	5/1 (83.33%)
Slap Happy	Joust	#103 Middleweight	Large mace-like hammer	1/1 (50.00%)
SlapPest	Team Revelation	#161 Middleweight		0/2 (0.00%)
Sledge-Jammer	Fat Cats	#79 Heavyweight	Rotating double-headed sledgehammer	1/2 (33.33%)
Slugger	Team Coolrobots	#130 Heavyweight	The Slugging Stick	0/1 (0.00%)
SMD	Team USAFA01	#83 Middleweight	Chainsaw	1/1 (50.00%)
Snake	Team Sinister	#51 Super Heavyweight	Big steel claws	1/1 (50.00%)
Snap Dragon	Team Gancarz	#85 Lightweight	Spring-loaded aluminum baseball bat	1/1 (50.00%)
Snipe	Invisible Works	#162 Lightweight	Rotational inertia	0/2 (0.00%)
SnowFlake	Team Toad	#33 Lightweight	Aluminum snowplow blade	3/2 (60.00%)
Snuggabot	Team United	#141 Middleweight	Horizontal 16-lb. 220 MPH spinning blade	0/1 (0.00%)
Some Parts	Raybotics	#93 Lightweight	Invertible wedge	1/1 (50.00%)
Son of Bob	Basher of Bots	#82 Middleweight		1/1 (50.00%)
Son of Body Parts	Team S.O.B	#195 Lightweight	Wedge	0/1 (0.00%)
Son of Smashy	Automatum	#14 Middleweight		5/0 (100%)
Son of Whyachi	Team Whyachi	#5 Super Heavyweight	60-in. 80-lb. 70 MPH spinning rotor	7/1 (87.50%)
Space Ape	X-plane Refugees	#74 Heavyweight	Vertical stainless steel rotary flail	1/1 (50.00%)
Space Madness	Obelisk	#95 Lightweight	Modular corner blades; 2-wheeled spinner; articulated conic wedge	1/1 (50.00%)
Space Monkey	X-plane Refugees	#27 Middleweight	2 vertical stainless steel rotary flails; modular attachment plate	4/1 (80.00%)
Space Operations Force	Robot Action League	#90 Middleweight		1/2 (33.33%)
Sparky	Team Sparky	#96 Lightweight	Vertical spinner	1/1 (50.00%)
Spaz	Team Vicious	#9 Middleweight	Hammer spike	6/2 (75.00%)
Speed Bump	BotBashers	#119 Lightweight	Wedge	0/2 (0.00%)
Speed Bump XL	BotBashers	#206 Middleweight	Drum may be electric powered	0/1 (0.00%)
Spike Demon		#124 Middleweight	Wedge shape	0/1 (0.00%)

TT	TM	KOs	KO%	AKOT	KOA	KOA%	AKOAT	TJD	JPA	AA	DA	SA	Points	Builder	Robot Name
2	6	2	33.33	0:02:02	2	33.33	0:01:12	0	0	0	0	0	8	Arndt Anderson	Shark Byte
1	1	0	0	0:00:00	0	0	0:00:00	1	6	2	2	2	0	David Chamberlain	Sharp Tooth
1	2	0	0	0:00:00	0	0	0:00:00	1	18	6	7	5	0	Raymond Roberts	She Fights Like a Girl
2	3	0	0	0:00:00	2	66.66	0:01:57	1	34	13	10	11	2	Richard Wood	Shish-ka-bot 1.1
2	2	0	0	0:00:00	0	0	0:00:00	2	16.5	5.5	4	7	0	James Arluck	Shockwave
1	1	0	0	0:00:00	0	0	0:00:00	1	4	2	0	2	0	Lorena Blalock	Short Circuit
2	5	1	20	0:01:09	1	20	0:01:28	3	25	8	8	9	6	Mark Curley	Short Order Chef
1	2	0	0	0:00:00	1	50	0:01:25	1	23	7	8	8	2	Joseph Murphy	Shovit
1	1	0	0	0:00:00	1	100	0:00:42	0	0	0	0	0	0	Larry Beckman	Shredbot
3	8	3	37.5	0:02:49	2	25	0:01:45	1	0	0	0	0	8	Steven Levin	Shrike
1	2	1	50	0:01:10	1	50	0:01:48	0	0	0	0	0	2	David Stevens	Silver Bullet
1	3	1	33.33	0:01:57	0	0	0:00:00	2	16.5	5	5.5	6	4	Robert Farrow	Silverback
1	4	1	25	0:01:50	1	25	0:01:43	2	28.5	9.5	9.5	9.5	6	Daniel Fernandez	Sisyphus
2	5	2	40	0:01:18	1	20	0:02:58	2	27.5	9.5	9	9	6	Robert Everhart	Skid Mark
2	7	1	14.28	0:01:38	0	0	0:00:00	6	27.33	8.33	9.33	9.67	10	Scott Kincaid	Slam Job
1	6	1	16.66	0:01:02	0	0	0:00:00	5	27.8	9.4	9.4	9	10	William Sauro	Slap 'Em Silly
1	2	0	0	0:00:00	0	0	0:00:00	1	14	3	3	8	2	Dave Owens	Slap Happy
1	1	0	0	0:00:00	0	0	0:00:00	0	0	0	0	0	0	Todd Mendenhall	SlapPest
2	3	0	0	0:00:00	2	66.66	0:01:01	0	0	0	0	0	2	William Sauro	Sledge-Jammer
1	1	0	0	0:00:00	1	100	0:00:54	0	0	0	0	0	0	Christian Carlberg	Slugger
1	2	1	50	0:01:47	1	50	0:01:20	0	0	0	0	0	2	Barry Mullins	SMD
1	2	1	50	0:02:12	1	50	0:02:26	0	0	0	0	0	2	Mark Setrakian	Snake
1	2	1	50	0:01:41	0	0	0:00:00	1	11	6	2	3	2	Gregory Gancarz	Snap Dragon
2	1	0	0	0:00:00	1	100	0:00:30	0	0	0	0	0	0	Charles Habermann	Snipe
2	5	2	40	0:01:06	1	20	0:02:11	2	20	7	8	5	6	Michael Mauldin	SnowFlake
1	1	0	0	0:00:00	0	0	0:00:00	1	8	2	3	3	0	Glenn Pipe	Snuggabot
1	2	0	0	0:00:00	0	0	0:00:00	2	28	9	10	9	2	Ramon Ebert	Some Parts
1	2	1	50	0:01:44	0	0	0:00:00	1	22	7	4	11	2	Wray Russ	Son Of Bob
1	1	0	0	0:00:00	1	100	0:00:57	0	0	0	0	0	0	Cuong Chau	Son of Body Parts
1	5	3	60	0:01:55	0	0	0:00:00	0	0	0	0	0	12	Derek Young	Son of Smashy
2	8	5	62.5	0:01:24	0	0	0:00:00	2	29	7	12.5	9.5	16	Terry Ewert	Son of Whyachi
1	2	0	0	0:00:00	1	50	0:00:54	1	37	11	12	14	2	Steven Bremer	Space Ape
1	2	0	0	0:00:00	0	0	0:00:00	2	25	8	9.5	7.5	2	Matt Lukes	Space Madness
1	5	1	20	0:00:37	1	20	0:01:32	3	24.67	8.33	7.67	8.67	8	Steven Bremer	Space Monkey
2	3	0	0	0:00:00	0	0	0:00:00	3	16.67	5	6	5.67	2	Michael Winter	Space Operations Force
1	2	0	0	0:00:00	0	0	0:00:00	2	25	8	9	8	2	Lee Kohn	Sparky
2	8	1	12.5	0:02:37	2	25	0:03:10	4	30.5	9.75	11	9.75	14	Mike Regan	Spaz
2	2	0	0	0:00:00	0	0	0:00:00	2	12	4	3.5	4.5	0	Joseph Murawski	Speed Bump
1	0	0	0	0:00:00	0	0	0:00:00	0	0	0	0	0	0	Joseph Murawski	Speed Bump XL
1	1	0	0	0:00:00	0	0	0:00:00	1	17	8	7	2	0		Spike Demon

Robot Name	Team	Ranking	Weapon	W/L (W/L%)
Spike of Doom	Team Hazard	#70 Lightweight		2/2 (50.00%)
Spin Cycle	Sqeeky Kleen	#150 Middleweight	Rotational inertia	0/1 (0.00%)
Spin Reaper	Spin Reaper	#108 Lightweight	24-lb. steel spinning ring with teeth	1/1 (50.00%)
Spinne	Archangel	#100 Heavyweight		0/1 (0.00%)
Spinning Mayhem	Pole Cat	#86 Heavyweight	Spinning blades and hammers	0/2 (0.00%)
Spinster	Team Deranged	#158 Middleweight	Spinning disk with blades	0/2 (0.00%)
Squirrelly D	BC Racing	#135 Middleweight	3600 RPM vertical lawn mower blade	0/1 (0.00%)
Stealth Terminator	Team USH	#29 Heavyweight	Three flipping sides	3/1 (75.00%)
Stealthbot — The Tallahassee Lassie	Area 52	#190 Middleweight	Spinning blade	0/1 (0.00%)
Steel Reign	Team XD	#68 Super Heavyweight	Mass; spikes; 2 triangle blades	0/2 (0.00%)
Stewbot	Stewbot	#196 Lightweight	The Stewicide Machine	0/1 (0.00%)
Stinger	Stinger	#52 Lightweight		2/2 (50.00%)
Stingray	Stingray	#92 Middleweight	Telescopic ramming spike	1/2 (33.33%)
Strange Brew	Team STUPID	#58 Middleweight	Lifting arm on a wedge	2/1 (66.67%)
Strike Terror	Team Mauser	#64 Super Heavyweight	24-in. spinning wheel with 4 teeth; rear skirt	1/1 (50.00%)
Stuffie	Stuffie	#84 Middleweight		2/2 (50.00%)
Subject to Change Without Reason	Team Boomer	#95 Middleweight	Wedge	1/3 (25.00%)
Sublime	Team Duct Tape	#109 Lightweight	4000 RPM steel-tipped titanium blade	1/2 (33.33%)
Suicidal Tendencies	Team Suicide	#68 Heavyweight	Pickax; wedge scoop	1/2 (33.33%)
Summoner	Robo Command	#28 Middleweight	Thick titanium front; small chisel-tipped saw blade	4/1 (80.00%)
Sunshine Lollibot	Team Sunshine	#22 Middleweight	24-in. 1800 RPM saw blade	4/2 (66.67%)
SupercaliBotulistic	Valhalla	#117 Lightweight	Flipping ramp	0/2 (0.00%)
Super Shark	Blood Sweat and Spears	#109 Super Heavyweight	Clamping jaws	0/1 (0.00%)
Surgeon General	Loki Robotics	#12 Heavyweight	Spinning blade	5/3 (62.50%)
The Swarm	The Swarm	#66 Super Heavyweight	Hive mentality; strength through numbers	0/2 (0.00%)
Swirlee	Pain USA	#11 Super Heavyweight	46-in. disk with 4-in. teeth, 86 MPH tooth-tip speed	5/1 (83.33%)
T.D.M	X-plane Refugees	#197 Lightweight	Outer spinning titanium shell	0/1 (0.00%)
T.U.S.K	Team Saber	#143 Lightweight	Speed kills	0/1 (0.00%)
T6	Team T6	#113 Middleweight	Wedge	0/1 (0.00%)
Tantrum	Demolition Team	#131 Lightweight	Lifting forks	0/1 (0.00%)
Tarkus	Manticore	#90 Heavyweight	Flipper head	0/1 (0.00%)
Tatar	Tatar	#142 Heavyweight	Spring loaded pike	0/1 (00.00%)
Tazbot	Mutant Robots	#6 Heavyweight	Twin-handled custom pickax	10/7 (58.82%)
Techno Destructo	Carnage	#8 Super Heavyweight	Pneumatic flipping arm	6/2 (75.00%)
Technofool	Team noFool	#72 Heavyweight	10-lb. hammers; 150-lb. 1000 RPM flywheel	1/3 (25.00%)
Tentoumushi 7.0	Robot Action League	#10 Lightweight	Grinder	7/6 (53.85%)
Terror	Team TNT2	#98 Heavyweight	10-in. pneumatic cylinder	0/1 (0.00%)
Tesla's Tornado	Team McHargue	#60 Middleweight	Spins	2/1 (66.67%)
Texas Stampede	Team Stampede	#132 Lightweight	Kinetic energy	0/1 (0.00%)

TT	TM	KOs	KO%	AKOT	KOA	KOA%	AKOAT	TJD	JPA	AA	DA	SA	Points	Builder	Robot Name
1	4	1	25	0:00:42	2	50	0:01:23	0	0	0	0	0	2		Spike of Doom
1	1	0	0	0:00:00	0	0	0:00:00	1	1	1	0	0	0	Dean Roberts	Spin Cycle
1	2	0	0	0:00:00	1	50	0:02:30	1	25	8	8	9	2	Ryan Hahn	Spin Reaper
1	1	0	0	0:00:00	0	0	0:00:00	1	7	3	2	2	0	Nathan Barnes	Spinne
2	1	0	0	0:00:00	0	0	0:00:00	1	19	4	6	9	0	Lowell Carpenter	Spinning Mayhem
2	2	0	0	0:00:00	2	100	0:01:08	0	0	0	0	0	0	Grayson DuRaine	Spinster
1	1	0	0	0:00:00	0	0	0:00:00	1	10	3	5	2	0	Patrick King	Squirrelly D
1	4	1	25	0:01:57	1	25	0:01:42	1	41	14	13	14	6	Rodger Farr	Stealth Terminator
1	1	0	0	0:00:00	1	100	0:01:09	0	0	0	0	0	0	Angelo Tsoukalas	Stealthbot - The Tallahassee Lassie
2	2	0	0	0:00:00	1	50	0:01:13	1	22	8	7	7	0	Brady Davis	Steel Reign
1	1	0	0	0:00:00	1	100	0:01:26	0	0	0	0	0	0	John Hargrave	Stewbot
1	4	1	25	0:00:54	2	50	0:04:21	0	0	0	0	0	4	Brad Geving	Stinger
2	3	0	0	0:00:00	0	0	0:00:00	3	15.67	6	5	4.67	2	Walter "Buck" McGibbony	Stingray
1	3	1	33.33	0:01:53	0	0	0:00:00	2	22.5	9	7.5	6	4	Dan L. Wiseman	Strange Brew
1	2	0	0	0:00:00	1	100	0:01:35	0	0	0	0	0	0	Richard Chandler	Strike Terror
1	4	1	25	0:04:40	1	25	0:03:25	0	0	0	0	0	2	Dave Chapman	Stuffie
3	4	0	0	0:00:00	2	50	0:01:59	2	21.5	6	8.5	7	2	Randy Eubanks	Subject to Change Without Reason
2	3	0	0	0:00:00	1	33.33	0:02:48	2	11.5	5	2.5	4	2	Jerome Miles	Sublime
2	3	1	33.33	0:02:03	0	0	0:00:00	2	7.5	7.5	0	0	2	Charles Binns	Suicidal Tendencies
1	5	1	20	0:00:41	0	0	0:00:00	4	23.75	7	8.75	8	8	Pierre Smith	Summoner
2	6	3	50	0:00:50	1	16.66	0:01:16	2	21	5	8.5	7.5	8	Andrew Miller	Sunshine Lollibot
2	2	0	0	0:00:00	0	0	0:00:00	2	18	7.5	5	5.5	0	Jay Johnson	SupercaliBotulistic
1	0	0	0	0:00:00	0	0	0:00:00	0	0	0	0	0	0		Super Shark
3	8	5	62.5	0:01:38	2	25	0:01:35	1	15	5	2	8	10	Eddy Ampuero	Surgeon General
2	2	0	0	0:00:00	0	0	0:00:00	2	21	10	7	4	0		Swarm, The
1	6	2	33.33	0:02:21	1	16.66	0:01:02	2	24	3.5	6.5	14	10	Kevin McEnery	Swirlee
1	1	0	0	0:00:00	1	100	0:00:44	0	0	0	0	0	0	Lester Knox	T.D.M.
1	1	0	0	0:00:00	0	0	0:00:00	1	13	3	4	6	0	John Patrick	T.U.S.K.
1	1	0	0	0:00:00	0	0	0:00:00	1	22	8	7	7	0	Ryan Bialaszewski	T6
1	1	0	0	0:00:00	0	0	0:00:00	1	18	5	6	7	0	Paul Ventimiglia	Tantrum
1	1	0	0	0:00:00	0	0	0:00:00	1	18	5	7	6	0	David Cain	Tarkus
1	0	0	0	0:00:00	0	0	0:00:00	0	0	0	0	0	0	Eric Stoliker	Tatar
6	17	2	11.76	0:02:56	2	11.76	0:03:15	11	23.91	7.55	7.82	8.55	21	Donald Hutson	Tazbot
2	8	0	0	0:00:00	2	25	0:02:26	6	28.17	9.5	9	9.67	12	Sean J. Irvin	Techno Destructo
3	2	0	0	0:00:00	0	0	0:00:00	2	19	5	6	8	2	Andrew Robinson	Technofool
5	13	3	23.07	0:02:58	3	23.07	0:02:41	5	22.6	8.6	5.2	8.8	14	Lisa Winter	Tentoumushi 7.0
1	1	0	0	0:00:00	0	0	0:00:00	1	8	2	1	5	0	Mark Talcott	Terror
1	3	1	33.33	0:02:16	0	0	0:00:00	1	6	0	1	5	4	William McHargue	Tesla's Tornado
1	1	0	0	0:00:00	0	0	0:00:00	1	18	7	5	6	0	Thomas Robertson	Texas Stampede

Robot Name	Team	Ranking	Weapon	W/L (W/L%)
That Good R&D	I'm That Good	#131 Heavyweight	Spinning blade	0/1 (0.00%)
Thorn	Loki Robotics	#198 Lightweight		0/1 (0.00%)
Thump	Team Digital	#124 Lightweight	Pickax	0/1 (0.00%)
Thumper	RedBaron	#132 Middleweight	Spinning steel bars	0/1 (0.00%)
Thwak	Team Vicious	#34 Middleweight	Vertical spinning weapon; wedge on the back	3/1 (75.00%)
Tiki Fire	Team Tiki	#199 Lightweight	Whacking arm	0/1 (0.00%)
Timber Wolf	Team Timber Wolf	#105 Super Heavyweight	High-powered motors	0/1 (0.00%)
Timmy	Truly Unruly	#33 Middleweight	Wedge; spinning weapon	3/2 (60.00%)
T-Minus	Inertia Labs	#3 Middleweight	Pneumatic flipper	7/2 (77.78%)
Tobor Rabies	Team Rabies	#144 Middleweight	Spikes; wings; saw	0/1 (0.00%)
Toe Crusher	Team Coolrobots	#11 Lightweight	Hammer; spike	7/5 (58.33%)
Tool Chop	Team Wizard	#92 Heavyweight	Linear actuator lift; spinning cutting wheel	0/1 (0.00%)
Toro	Inertia Labs	#3 Super Heavyweight	Pneumatic flipper	9/2 (81.82%)
Torquemada	Warhippies	#76 Super Heavyweight	Cow catcher; teeth	0/1 (0.00%)
Tortoise	Achilles	#48 Super Heavyweight	Kinetic shield with 0.5-in. steel slabs	1/3 (25.00%)
Total Chaos	Team Total Chaos	#132 Heavyweight	4-sided wedge	0/1 (0.00%)
Tower of Power	Tower of Power	#114 Middleweight	7-ft. forklift	0/1 (0.00%)
Towering Inferno	The Infernolab	#40 Heavyweight	Big hammers	2/2 (50.00%)
Trainwreck	ICR	#136 Middleweight	Kinetic ram	0/1 (0.00%)
Travesty	Team Travesty	#76 Middleweight	Pusher	1/1 (50.00%)
Traxx	Restless Nights	#66 Middleweight	Scoop; hammer	2/2 (50.00%)
Trial and Error	Team Trial and Error	#86 Super Heavyweight	Twin triangular spinning steel drums	0/1 (0.00%)
TriDent	C2 Robotics	#65 Middleweight	Geared lifting arm	2/1 (66.67%)
Trig Terror	Team Malicious	#191 Middleweight	Saws	0/1 (0.00%)
Trilobot	Lungfish Technologies	#24 Lightweight	Muffler cut-off chisels	4/2 (66.67%)
TriMangle	Loki Robotics	#28 Super Heavyweight	Spinning triangle; flipper; bulldozer	2/2 (50.00%)
Tripulta Raptor	Team Raptor	#66 Heavyweight	Hydraulic jaws	1/2 (33.33%)
Tripulta Raptor	Team Raptor	#64 Super Heavyweight	Hydraulic jaws	1/1 (50.00%)
Turbo	Loki Robotics	#91 Middleweight	Sweet Knockers	1/3 (25.00%)
Turtle Road Kill	Team Attitude	#11 Middleweight		8/4 (66.67%)
Twin Paradox	Skunk-Tek	#6 Middleweight	Horizontal spinning disk	7/2 (77.78%)
Twist of Fate	Team Twist of Fate	#209 Lightweight	Dual flipping/lifting arms that turn robots over sideways	0/1 (0.00%)
Twister	Team Brute Force	#129 Middleweight	Rotational inertia	0/1 (0.00%)
T-Wrex	Regan Designs	#12 Middleweight	Ramming spikes and tail	6/2 (75.00%)
UGV Scorpion		#39 Lightweight	Spiked wedge with spiked stops	3/1 (75.00%)
Undertow	Team Undertow	#200 Lightweight	Spinning ax head; kinetic energy	0/1 (0.00%)
Unlicensed Chiropractor	Team In-Theory	#201 Lightweight	Assorted rams	0/1 (0.00%)
Uplifting Experience	Snowhite	#63 Middleweight	Lifter	2/1 (66.67%)
Velocity	Team Velocity	#125 Middleweight	Ramming and pushing	0/2 (0.00%)
Vermicous Kenid	Precision Engineering	#60 Lightweight	Flying Wedge/Wheel of Misfortune	2/2 (50.00%)

TT	TM	KOs	KO%	AKOT	KOA	KOA%	AKOAT	TJD	JPA	AA	DA	SA	Points	Builder	Robot Name
1	1	0	0	0:00:00	1	100	0:01:53	0	0	0	0	0	0	Jason Stokes	That Good R&D
1	1	0	0	0:00:00	1	100	0:02:35	0	0	0	0	0	0	Korey Kline	Thorn
1	1	0	0	0:00:00	0	0	0:00:00	1	22	9	8	5	0	Chad Jackson	Thump
1	1	0	0	0:00:00	0	0	0:00:00	1	12	4	6	2	0	Scott Baron	Thumper
1	4	1	25	0:00:42	1	25	0:01:37	2	29.5	8.5	12	9	6	Mike Regan	Thwak
1	1	0	0	0:00:00	1	100	0:01:29	0	0	0	0	0	0	Micah Leibowitz	Tiki fire
1	1	0	0	0:00:00	1	100	0:00:30	0	0	0	0	0	0	John Tuminaro	Timber Wolf
2	5	2	40	0:01:52	2	40	0:01:01	0	0	0	0	0	6	John Waldron	Timmy
2	9	7	77.77	0:01:19	1	11.11	0:05:20	0	0	0	0	0	16	Reason Bradley	T-Minus
1	1	0	0	0:00:00	0	0	0:00:00	1	7	1	3	3	0		Tobor Rabies
4	12	3	25	0:02:44	2	16.66	0:03:06	3	33.33	10.67	11.67	11	13	Christian Carlberg	Toe Crusher
1	1	0	0	0:00:00	0	0	0:00:00	1	17	6	5	6	0	Luis Floriano	Tool Chop
3	11	4	36.36	0:00:59	1	9.09	0:02:35	6	33.5	11.17	10.33	12	24	Alexander Rose	Toro
1	1	0	0	0:00:00	0	0	0:00:00	1	17	8	4	5	0	Creighton Nachtigall	Torquemada
3	4	1	25	0:01:04	1	25	0:00:54	2	2.5	0.5	0	2	2	Ian Burt	Tortoise
1	1	0	0	0:00:00	1	100	0:01:21	0	0	0	0	0	0	Mike Hinson	Total Chaos
1	1	0	0	0:00:00	0	0	0:00:00	1	22	8	8	6	0	Dan Haeg	Tower Of Power
2	4	1	25	0:01:17	2	50	0:01:20	1	23	9	8	6	4	Jason Bardis	Towering Inferno
1	1	0	0	0:00:00	0	0	0:00:00	1	10	4	2	4	0	Joe Richards	Trainwreck
1	2	1	50	0:01:00	0	0	0:00:00	1	19	7	6	6	2	Doug Jones	Travesty
2	3	0	0	0:00:00	0	0	0:00:00	3	20	7.33	6	6.67	4	Daniel Goff	Traxx
1	1	0	0	0:00:00	0	0	0:00:00	1	1	1	0	0	0	Ron Whitton	Trial & Error
1	3	0	0	0:00:00	0	0	0:00:00	3	29.33	11	9.33	9	4	Luke Khanlian	TriDent
1	1	0	0	0:00:00	1	100	0:02:16	0	0	0	0	0	0	Daniel Rupert	Trig Terror
2	6	2	33.33	0:01:09	2	33.33	0:02:25	2	25	7.5	9.5	8	8	Phil Putman	Trilobot
2	4	1	25	0:00:41	0	0	0:00:00	3	17.67	5.33	5.67	6.67	4	Korey Kline	TriMangle
2	3	1	33.33	0:01:43	1	33.33	0:02:36	0	0	0	0	0	2	Chuck Pitzer	Tripulta Raptor
1	2	0	0	0:00:00	1	50	0:00:55	0	0	0	0	0	2	Chuck Pitzer	Tripulta Raptor
3	4	0	0	0:00:00	1	25	0:03:49	2	23.5	7	10.5	6	2	Eddy Ampuero	Turbo
3	12	4	33.33	0:01:15	2	16.66	0:01:43	4	17.25	6	5	6.25	12	Sam Steyer	Turtle Road Kill
2	9	4	44.44	0:01:15	1	11.11	0:01:43	4	21.75	7	7.25	7.5	14	Michael Neese	Twin Paradox
0	0	0	0	0:00:00	0	0	0:00:00	0	0	0	0	0	0	Alan Thompson	Twist of Fate
1	1	0	0	0:00:00	0	0	0:00:00	1	13	4	5	4	0	Charles Steinkuehler	Twister
2	8	4	50	0:01:45	0	0	0:00:00	3	25.67	7.67	9.33	8.67	12	Brent Regan	T-Wrex
1	4	1	25	0:02:26	0	0	0:00:00	3	26.67	9.67	8.33	8.67	6	Brett Dawson	UGV Scorpion
1	1	0	0	0:00:00	1	100	0:01:23	0	0	0	0	0	0	Bryan Boucher	Undertow
1	1	0	0	0:00:00	1	100	0:01:13	0	0	0	0	0	0	Will Tatman	Unliscensed Chiropractor
1	3	0	0	0:00:00	0	0	0:00:00	3	31.67	10.67	10.33	10.67	4	Bob Schneeveis	Uplifting Experience
2	2	0	0	0:00:00	1	50	0:00:53	1	13	4	7	2	0	Steve Yoder	Velocity
2	4	1	25	0:01:44	0	0	0:00:00	3	16.67	4.67	6	6	4	Jared Sexty	Vermicous Kenid

Robot Name	Team	Ranking	Weapon	W/L (W/L%)
Viking Dragon	Team Viking USA	#192 Middleweight	Rebar and steel spikes	0/1 (0.00%)
Village Idiot	Team RV	#23 Middleweight	Dual 12-in. saw blades	4/2 (66.67%)
Violator	TuxBot	#48 Middleweight	CO_2-powered spike	2/2 (50.00%)
Vlad the Impaler	Vladmeisters	#2 Heavyweight	Lifting spears	16/4 (80.00%)
Vladiator	Vladmeisters	#4 Super Heavyweight	Lifting spear with can-opener feature	8/1 (88.89%)
Voltronic	Voltronic	#3 Heavyweight	Articulated lifting arm	12/7 (63.16%)
Vortex	Team Vortex	#193 Middleweight	5-lb. trench-digging tool; auxiliary wedge	0/1 (0.00%)
W.L.O.W	Team Raptor	#68 Lightweight		3/2 (60.00%)
Wack-a-tron	Team Wack-a-tron	#106 Super Heavyweight	60-lb. 40.5-in 200 RPM circular saw blade	0/1 (0.00%)
Wacker	Snowhite	#37 Lightweight	Rotating blades	3/1 (75.00%)
The Wacky Compass	Whacky Compass	#80 Lightweight		1/1 (50.00%)
War Machine	Team Delta	#24 Super Heavyweight	Wedge	3/1 (75.00%)
Wedge of Doom	Team Hazard	#7 Lightweight	Actuated lifting arm	8/2 (80.00%)
Wedge-O-Matic	Bot Boyz	#57 Lightweight	Wedge	2/2 (50.00%)
Weed Wacker	Screwed	#202 Lightweight		0/1 (0.00%)
Wham! Bam!	TeamPeril	#100 Lightweight	Bat	1/2 (33.33%)
Whirl Wep — EEL Brain	Robot Action League	#53 Lightweight	Whirling disk	2/2 (50.00%)
Whirligig	Team Sorcerer's Apprentice	#18 Lightweight	Spiked hammer arm; cyclone action	5/2 (71.43%)
Whirling Willy	Destructor Constructors	#78 Lightweight	Momentum	1/1 (50.00%)
Why Not	Team Whyachi	#194 Middleweight	Triangular horizontal spinner	0/1 (0.00%)
Whyachi	Team Whyachi	#61 Super Heavyweight	50 MPH spinning steel hammers	1/1 (50.00%)
Widowmaker	Widowmaker	#74 Lightweight	Spinning bar	1/1 (50.00%)
WindChill	Team Toad	#88 Middleweight	Steel ramming spike	1/2 (33.33%)
World Peace	Team Bohica	#26 Super Heavyweight	The BEAK	2/5 (28.57%)
WulfBane	APSF Robotics Division	#75 Middleweight	Double-bladed battle ax	1/1 (50.00%)
Yabba Dabba Dozer	MITGIT	#77 Super Heavyweight	Spinning steel drum; bulldozer blade	0/1 (0.00%)
Yoshik	Team Tekken	#134 Heavyweight		0/1 (0.00%)
Yoshikillerbot	Lightning	#195 Middleweight	Front wedge; spinning dual-edged ax	0/1 (0.00%)
Y-Pout	Team Whyachi	#133 Heavyweight	Triangular horizontal spinner	0/1 (0.00%)
YU812	Team Whyachi	#71 Lightweight	Triangular horizontal spinner	1/1 (50.00%)
Ziggo	Team Ziggy	#1 Lightweight	Tool steel blades, 15,000 joules	18/3 (85.71%)
ZiggZagg	Team Ziggy	#52 Middleweight	Front lifting arm	2/1 (66.67%)
Zion	Zion	#8 Middleweight	Wedge; spikes	7/2 (77.78%)
Zombie	Necrobotics	#196 Middleweight	Sword; axial blade array; ramming spike; plough	0/1 (0.00%)
Zugzwang	Team Entropy	#203 Lightweight	3 spinning disks	0/1 (0.00%)
Zulu	DreamDroid	#197 Middleweight	Stabbing arm	0/1 (0.00%)

TT	TM	KOs	KO%	AKOT	KOA	KOA%	AKOAT	TJD	JPA	AA	DA	SA	Points	Builder	Robot Name
1	1	0	0	0:00:00	1	100	0:01:28	0	0	0	0	0	0	Eric Dalnes	Viking Dragon
2	6	3	50	0:01:11	1	16.66	0:01:35	2	22.5	9	9	4.5	8	Jon Autry	Village Idiot
2	4	1	25	0:00:42	1	25	0:01:50	1	14	5	3	6	4	Eric Molitor	Violator
5	20	4	20	0:01:41	2	10	0:01:35	10	36.2	11.7	11.9	12.6	37	Gage Cauchois	Vlad the Impaler
2	9	1	11.11	0:02:03	0	0	0:00:00	8	26.88	9.5	9.63	7.75	18	Gage Cauchois	Vladiator
6	19	2	10.52	0:01:14	3	15.78	0:02:51	13	30.31	9.77	10.85	9.69	26	Stephen Felk	Voltronic
1	1	0	0	0:00:00	1	100	0:01:40	0	0	0	0	0	0	Matt Hockenheimer	Vortex
1	5	3	60	0:01:39	0	0	0:00:00	0	0	0	0	0	3		W.L.O.W.
1	1	0	0	0:00:00	1	100	0:00:19	0	0	0	0	0	0	Dennis Shellhouse	Wack-a-tron
1	4	1	25	0:01:20	1	25	0:01:28	2	35.5	11	13	11.5	6	Bob Schneeveis	Wacker
1	2	1	50	0:01:24	1	50	0:01:28	0	0	0	0	0	2	Christopher Baron	Wacky Compass, The
1	4	1	25	0:02:26	1	25	0:01:25	2	41.5	14.5	12.5	14.5	6	Dan Danknick	War Machine
2	10	4	40	0:01:33	1	10	0:02:15	5	29.6	10.2	9.4	10	16	Tony Buchignani	Wedge of Doom
2	4	1	25	0:01:23	1	25	0:02:24	2	17.5	6	6	5.5	4	Tim Anderson	Wedge-O-Matic
1	1	0	0	0:00:00	1	100	0:01:08	0	0	0	0	0	0	Olivier Musy-Verdel	Weed Wacker
2	3	0	0	0:00:00	1	33.33	0:00:48	2	22.5	8	8.5	6	2	Bruce Eddie Switzer	Wham! Bam!
2	4	1	25	0:01:00	1	25	0:01:24	2	24	6.5	10	7.5	4	Michael Winter	Whirl Wep – EEL Brain
2	7	2	28.57	0:02:30	2	28.57	0:01:47	2	30	8	11	11	10	Jeremy Franklin	Whirligig
1	2	1	50	0:01:10	1	50	0:01:32	0	0	0	0	0	2	Keith Van Sickle	Whirling Willy
1	1	0	0	0:00:00	1	100	0:00:44	0	0	0	0	0	0	Terry Ewert	Why Not
1	2	0	0	0:00:00	0	0	0:00:00	1	17	5	6	6	2	Terry Ewert	Whyachi
1	2	1	50	0:01:04	0	0	0:00:00	1	19	8	6	5	2	Kevin Lillie	Widowmaker
2	3	0	0	0:00:00	0	0	0:00:00	3	18.33	7	6	5.33	2	Michael Mauldin	WindChill
5	6	0	0	0:00:00	2	33.33	0:01:35	3	24	6	8.33	9.67	6	Dave Campbell	World Peace
1	2	1	50	0:00:55	0	0	0:00:00	1	12	5	2	5	2	Michael Jacobson	WulfBane
1	1	0	0	0:00:00	0	0	0:00:00	1	16	5	4	7	0	William Garcia	Yabba Dabba Dozer
1	1	0	0	0:00:00	1	100	0:01:39	0	0	0	0	0	0	Steven Karp	Yoshik
1	1	0	0	0:00:00	1	100	0:00:46	0	0	0	0	0	0	Steven Karp	Yoshikillerbot
1	1	0	0	0:00:00	1	100	0:01:02	0	0	0	0	0	0	Terry Ewert	Y-Pout
1	2	1	50	0:00:44	0	0	0:00:00	1	22	4	6	12	2	Terry Ewert	YU812
5	21	16	76.19	0:01:48	2	9.52	0:00:58	1	24	6	8	10	43	Jonathan Ridder	Ziggo
1	3	1	33.33	0:00:51	0	0	0:00:00	2	17	5.5	6.5	5	4	Jonathan Ridder	ZiggZagg
2	9	2	22.22	0:02:44	0	0	0:00:00	6	26.5	9	8.5	9	14	Jered Singleton	Zion
1	1	0	0	0:00:00	1	100	0:00:55	0	0	0	0	0	0	Murray Finegold	Zombie
1	1	0	0	0:00:00	1	100	0:01:10	0	0	0	0	0	0	Chris Connor	Zugzwang
1	1	0	0	0:00:00	1	100	0:01:09	0	0	0	0	0	0	Walter Martinez	Zulu

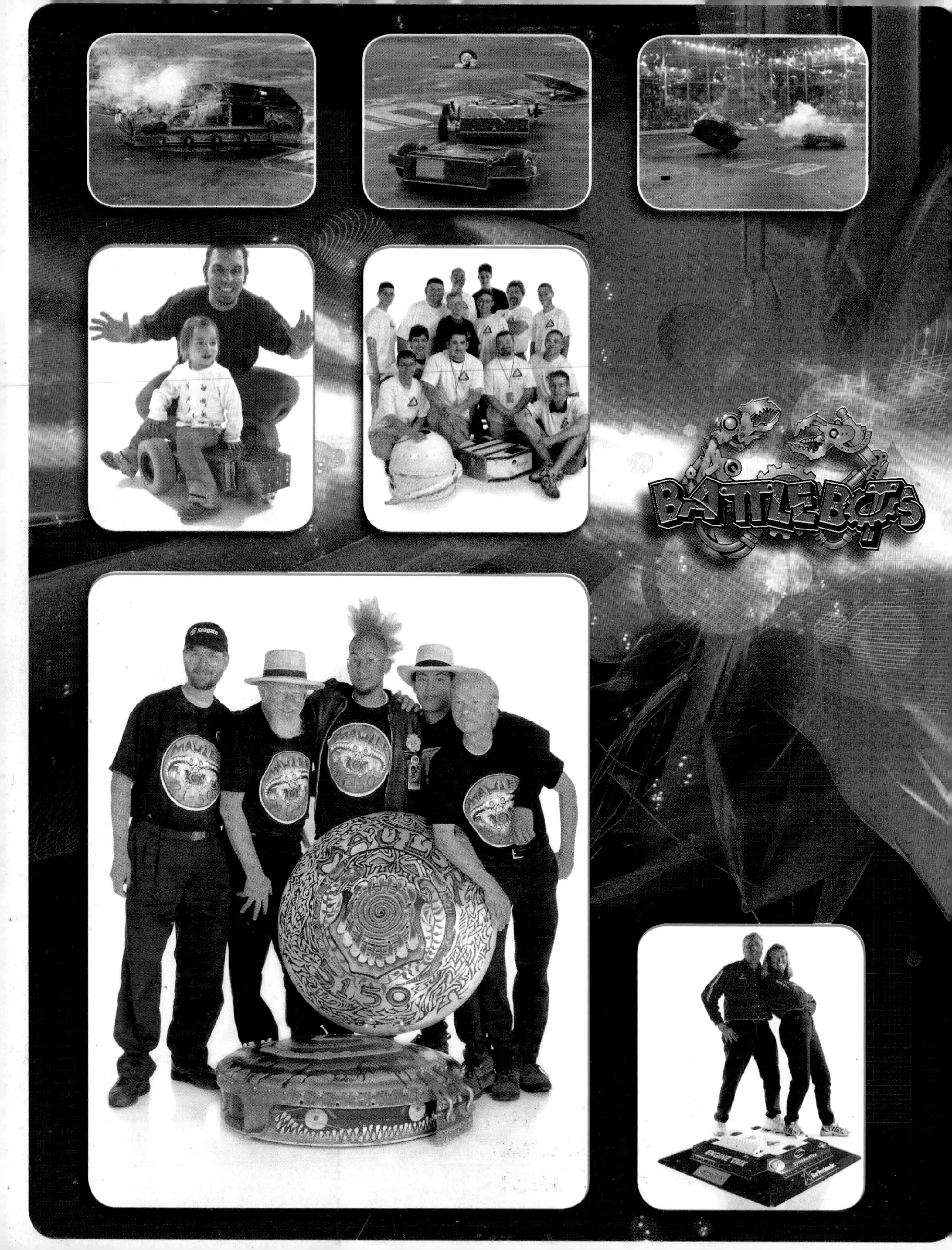
BATTLEBOTS
Seagate